GARDENING IN A CHANGING WORLD

Plants, People and the Climate Crisis

DARRYL MOORE

Pimpernel Press Ltd
www.pimpernelpress.com

Pimpernel Press Limited
www.pimpernelpress.com

A catalogue record for this book is available from the British Library.

ISBN 978-1-910258-28-6

Typeset in PT Sans and PT Serif
Printed and bound by CPI Group (UK) Ltd, Croydon CR0 4YY

9 8 7 6 5 4 3 2 1

CONTENTS

PREFACE

'I'VE BEEN LIVING THROUGH CHANGES,
IT'S NOT THE SAME THING EVERY DAY'
The Teardrop Explodes, 'Treason'

The ironic expression 'May you live in interesting times' has never seemed more apt, faced as we are with what appears to be an unpredictable world changing rapidly before our eyes. We are challenged not only by the climate crisis and biodiversity loss but also by global pandemics, all of which threaten to substantially disrupt and jeopardize both our lives and all the more-than-human life forms that surround us.

It is clear that the relationship between humans and the rest of nature has never before been so disconnected. Yet it is towards 'nature-based solutions' that organizations and governments are beginning to turn their attention. We are part of nature so this shouldn't be surprising, but unless we seriously reconsider our instrumental attitude towards it, we will either continue to make the same mistakes or create new problems. The British nature writer Richard Mabey has always been ambivalent about the idea of expecting the rest of nature to cure us of our maladies and troubles, as he rightly points out that many of them are caused in the first place by our neglect, disregard and maltreatment of it.

People look towards green spaces for some kind of 'nature cure', as they are some of the most common daily encounters many of us have with natural processes. In an increasingly post-industrial, urbanized world, gardens and parks offer opportunities for engagement that are beneficial for physical and mental health. This has become acutely apparent during Covid-19 lockdowns. Perhaps it shouldn't be surprising, as these are places with plants, evolutionary ancestors that have successfully lived on the planet for 420 million years. They know a thing or two about adapting and surviving through troubled times.

So perhaps we can learn from them, if we can first get to understand them better and move beyond many current attitudes towards them. In the history of gardens and parks, plants have been viewed primarily from an aesthetic perspective, and traditional horticultural habits have been based upon practices that are damaging to the environment and unsustainable. So, what would it be like if our engagement with plants was ecological instead? This would entail rethinking the ways we have worked and lived with them in the past, appreciating new approaches being used today, and giving consideration to how we may do so in the future by looking to wide-ranging knowledge sources from traditional indigenous cultures to contemporary science.

We should be asking ourselves: what would it be like to confront the challenges of a changing world if we approached plants not by simply using them as solutions, but by working with them as allies or kin in entangled networks with other species? We should be asking ourselves: what would it be like to face the future from a plant-centric perspective?

INTRODUCTION: A CHANGING WORLD

Planet Earth is a dynamic, changing place that has endured many dramatic geological and climatic events over its 4.5 billion year history. These have had extreme effects on the life forms that inhabit the Earth. The interaction of biotic and abiotic forces has shaped the mineral, aquatic and atmospheric realms of the place we call home, in an ongoing process of co-evolution over time.

But we are now living in a period in which the actions of one species are shaping the present and the future. Our species, *Homo sapiens*, has evolved from being one among many on the planet to become the dominant one that has dramatically changed the environments we live in. The development of consciousness, collective interaction, and the use of tools eventually led to the anthropocentric conceit that humans exist at a remove from the natural world they inhabit. Over time the binary distinction between nature and culture has produced a false sense of superiority and infallibility that has led *Homo sapiens* to become an animal who wants more than it needs.

As a consequence, we have begun to seriously affect the planet at a global scale through both intentional and unconscious actions and patterns of behaviour, creating novel geological strata, changing ecological systems, and altering the atmosphere to such a degree that we are now experiencing negative effects. These are manifesting themselves at an unprecedented pace and are set to continue and increase in the future. The challenges we face not only relate to our species but have repercussions for all the other life forms we are intricately, and often invisibly, interconnected with. We are having to confront the responsibilities of how we can continue to live with these life forms on a planet with finite resources. We are also having to face up to the fact that both the causes and the effects of the situation are not proportionally distributed across the globe. They are the product of uneven and unjust geopolitical development over many centuries, carried out by particular social groups, nations and corporations, and as a result those least culpable are likely to be the most affected. The culmination of all these factors is that we find ourselves in a novel situation of a rapidly changing world that we have unwittingly created, in a period that has fittingly been dubbed the Anthropocene.

The term was popularized by Dutch chemist Paul Crutzen, who was awarded the 1995 Nobel Prize in Chemistry for proving that the ozone layer, which shields the planet from ultraviolet light, was thinning at the poles because of rising concentrations of industrial gases. At a meeting of scientists who studied the planet's oceans, land surfaces and atmosphere in Mexico in 2000, he proposed it as a term for a new geological epoch succeeding the current one, the Holocene, which defines the past ten to twelve thousand postglacial years, as proposed at the International Geological Congress in Berlin in 1885. Crutzen argued that, as humans have inhabited or visited all places on Earth and their activities have become a growing geological and morphological force, which has been accelerating since the industrial period, there was a need to officially recognize this and its consequences.

Subsequently, a geological working group tasked with determining if there is enough evidence for the idea to meet the classification requirements for epochs defined the period as exhibiting phenomena that include

> an order-of-magnitude increase in erosion and sediment transport associated with urbanization and agriculture; marked and abrupt anthropogenic perturbations of the cycles of elements such as carbon, nitrogen, phosphorus, and various metals together with new chemical compounds; environmental changes generated by these perturbations, including global warming, sea-level rise, ocean acidification, and spreading oceanic 'dead zones'; rapid changes in the biosphere both on land and in the sea, as a result of habitat loss, predation, explosion of domestic animal populations and species invasions; and the proliferation and global dispersion of many new 'minerals' and 'rocks' including concrete, fly ash and plastics, and the myriad 'technofossils' produced from these and other materials.

And if that doesn't sound overwhelming enough, they go on to state, 'Many of these changes will persist for millennia or longer, and are altering the trajectory of the Earth System, some with permanent effect. They are being reflected in a distinctive body of geological strata now accumulating, with potential to be preserved into the far future.'[1]

Crutzen defined the Anthropocene as beginning around 1800 with the onset of industrialization, which saw an exponential expansion in the extraction and use of fossil fuels, but ongoing debates have sought to determine a definitive date, with the general consensus currently suggesting 1950 as a result of the peak in radioactive fallout of the nascent nuclear era. The term itself has also been contested for overemphasizing the importance of humans, although it is the negative influence of humans that it clearly highlights and questions. An alternative name, the Capitalocene, has been proposed by environmental historian Jason Moore, given the correlation between the dates of the observed effects of anthropocentric activities and the birth of industrial capitalism in the Global North.[2]

The phenomena described by the working group reveal a seriously disconnected relationship between humans and the rest of nature and natural processes. The most generally discernible and devastating evidence of the Anthropocene can clearly be seen in the effects of climate change, large-scale habitat loss and loss of biodiversity. Significant signs of climate change are increasingly revealing themselves in unpredictable seasonal global weather patterns and extreme weather events, made manifest in various ways, including rising temperatures, melting glaciers and sea ice, global mean sea-level rise and species migrations.[3] Devastating floods, droughts, hurricanes and wildfires are becoming common features on the news with a disconcerting frequency. In the past, the planet has undergone various climatic changes, enduring ice ages of long durations followed by more temperate periods. The Holocene saw relatively stable temperatures with deviations of only 1°C/1.8°F above or below the average. However, the rising temperatures of the Anthropocene point to overwhelming evidence that this is not part of some natural cycle, but that it is anthropogenically driven, caused by the actions of humans.

The primary driver is an increase in greenhouse gases, such as carbon dioxide (CO_2), methane (CH_4) and nitrous oxide (N_2O), which have always played an important role in the atmosphere, where their concentrations for thousands of years have remained within a fairly stable range, balanced out by natural absorption, allowing life to flourish. The principle of the 'greenhouse effect' was proposed by French mathematician Joseph Fourier in 1824 to describe the warming that happens when gases trap heat in the Earth's atmosphere. These gases let in light and heat from the sun, some of which is absorbed and some of which reflects back through the atmosphere. Some heat passes straight through and some is kept from escaping by the gases, like the glass walls of a greenhouse, maintaining a liveable temperature, without which the Earth's surface would be an average of about 30°C/86°F colder.

In 1895 the Swedish chemist Svante Arrhenius discovered that burning coal could enhance the greenhouse effect by making carbon dioxide more prevalent in the atmosphere, thereby trapping more of the escaping heat.[4] Since then a significant increase in greenhouse gas emissions has enhanced the greenhouse effect, discernibly warming the Earth. This leap corresponds with a period of intensified fossil fuel extraction and consumption, net land change use, including mass deforestation for agriculture (half the world's tropical forests have been lost and continue to disappear at a rate of 12 million hectares/30 million acres a year), and, since the mid-twentieth century, the rapid spread of global capitalism and post-industrial consumerism.

Greenhouse gas concentrations of carbon dioxide, methane and nitrous oxide in the atmosphere have increased to levels unprecedented in at least the last 800,000 years, with a slow but steady rise since 1750 and a discernible leap since 1960.[5] Over 365 billion metric tons of carbon have been added to the atmosphere from fossil fuel emissions, with another 180 billion tons from deforestation. In October 2021 the concentrations of the greenhouse gases above were 413 ppm, 1,907 ppb, and 334 ppb,[6] respectively, exceeding the pre-industrial levels by about 48 per cent, 172 per cent, and 23 per cent.[7] Predictions suggest that by 2050 CO_2 levels will be more than double that of the pre-industrial period and the highest they have been for 50 million years.[8]

The vast global oceans, covering 70 per cent of the planet's surface, have an enormous capacity for absorbing CO_2, taking up about a third of emissions by humans since industrial times.[9] But in the process of absorbing this extra CO_2, the oceans have experienced a change in pH level, meaning that they are now 30 per cent more acidic than they were in 1800.[10] As a consequence, global bleaching of coral reefs has become a major problem. Bleaching is a stress response that causes coral animals to expel the microscopic algae (zooxanthellae) whose photosynthesis provides the energy needed to build three-dimensional reef structures. Predictions suggest that, under a business-as-usual emissions scenario, all twenty-nine coral-containing World Heritage sites will cease to exist as functioning coral reef ecosystems by the end of this century.[11]

The oceans also absorb heat, meaning that an enormous amount of energy is required to raise the Earth's surface temperature by a small amount. But despite the fact that the oceans have absorbed more than 90 per cent of the extra heat we have been producing, the global annual temperature has still increased at an average rate of 0.08°C/0.14°F per decade since 1880 and over twice that rate (0.18°C/0.32°F) since 1981. The seven warmest years in the 1880 to 2020 records have all occurred since 2015, while the ten warmest years have occurred since 2005.[12]

The 1°C/2°F increase in global average surface temperature that has occurred since the pre-industrial era represents a significant increase in accumulated heat. The effects of this increase are regional and seasonal temperature extremes, reducing snow cover and melting sea ice, intensifying heavy rainfall, disrupting highly biodiverse ecosystems such as coral reefs and rainforests, and changing habitat ranges for plants and animals by expanding some and shrinking others.

Warming temperatures are causing many species to migrate northwards and to higher altitudes, but this is not happening with all species of a particular habitat or at the same pace. Some, generally the most mobile, are responding quickly; others, including trees and other plants, will only be able to migrate through reproduction and seed distribution. This is a problem, as it disrupts the ecological relationships at play in environments. Some species are dependent upon others for food and shelter, and as a result of uneven migration they may become separated, leaving some isolated populations to decline or go extinct. This is a particular concern for species in mountain habitats, which can only migrate upwards until they can go no further. This accelerated trend of migration and loss of habitats has become known as the sixth mass extinction event.

A series of dramatic events in the past have drastically transformed life on the planet. The Ordovician–Silurian, Late Devonian, Permian–Triassic, Triassic–Jurassic and Cretaceous–Paleogene periods all experienced mass extinction events, characterized by the loss of at least 75 per cent of species within a geologically short period of time. Unlike the previous extinctions, the current situation is seen as the result of anthropocentric climate change, aided and abetted by other related human activities such as deforestation, pollution, urbanization, the spread of agriculture and the prevalence of monocultures.

These causes are interlinked in many different ways and together make a toxic cocktail. Land use change and deforestation, largely for agricultural purposes such as feedstocks for industrially farmed animals to feed an expanding global population increasingly eating a 'Western diet', displace communities and erase cultural traditions, and bring humans into closer contact with wild species, causing spillover events of zoonotic viruses. Industrial farming brings with it soil degradation and disruption of hydrologic cycles, which can threaten food security and thereby human health. While the rationale behind these activities may seem sensible at an abstract level in terms of providing food, the reality of their negative consequences puts the balance of their benefits into question. Their dependency upon fossil fuels makes them major contributors to CO_2 production, and the ecological cost extends further to habitat loss and ecosystem destruction.

Habitats change as part of evolutionary processes but, as with rising temperatures, it is the rate at which change is now happening and the uneven effects that it is producing which are concerning. Habitats are not simply geographical locations; they are spatial arrangements that provide species with the resources and shelter necessary for their survival. They are important parts of ecosystems: the communities of different living species interacting with each other and the non-living elements of their environments. These relationships evolve over long periods of time, creating relative degrees of stability within an ecosystem. Dramatic fluctuations of species or disturbances to habitats have knock-on effects for many of the species within them by setting off positive feedback cycles, making the problem worse. If an ecosystem is not resistant to the changes, or resilient enough to bounce back from them, then there can be drastic consequences for many of the species within them. Loss of biodiversity undermines the productivity and effective functioning of ecosystems, many of which are essential not only to the species that inhabit them but also for many human activities.

The current extinction event is affecting species across the biological spectrum, including plants, mammals, birds, reptiles, amphibians, fish and invertebrates. While the extinction of many species has been documented, a great many have not been, as they are undiscovered at the time of their loss. But just because we are unaware of them doesn't mean they are unimportant, as we have yet to fully understand the ways in which they may affect the larger ecosystem networks they are part of. According to the International Union for Conservation of Nature (IUCN) Red Lists that record threatened species, more than 32,000 species face extinction, including, of all those evaluated, 41 per cent of amphibians, 26 per cent of mammals, 21 per cent of reptiles, 13 per cent of birds and, although there is currently insufficient coverage in the data, an estimated 40 per cent of plants.[13] Estimated current rates of extinction of species vary between a hundred to a thousand times higher than natural background rates, and scientists predict that more than one million species are on track for extinction in the coming decades.

The loss of plants is particularly significant, not only for their central role in ecosystems and food chains, but also because of the vital function they play in the carbon cycle through the process of photosynthesis and respiration. Terrestrial plants remove about 29 per cent of emissions that would otherwise contribute to growth of the atmospheric CO_2 concentration. They are worthy of greater attention based on that fact alone. As they are also essential for life on the planet as we know it, a better understanding and appreciation of them is essential. We must re-evaluate our relationship with plants.

PLANTS AS PRODUCERS

IN PRAISE OF PLANTS

Plants are all around us in a multitude of guises and sizes. Since their evolutionary migration out of aquatic habitats on to land 420 million years ago, plants have become adept at adapting to their surroundings, finding appropriate locations with compatible neighbours, and living in communities that allow them to flourish. Many of them, such as cycads, are geological time travellers that have successfully survived across eras, to be here still in more or less the same form as they were in the Permian period, 280 million years ago.

Plants have made their homes in nearly every terrestrial environment across the globe, and many aquatic ones, in some of the most inhospitable and unlikely places. They cover 75 per cent of the planet's terrestrial surface and constitute 80 per cent of its total biomass, while bacteria in all ecosystems are next, accounting for just 15 per cent.[1] They weigh in at an impressively hefty 450 gigaton of carbon, compared to just 2 gigaton of animals. There are approximately 390,000 plant species; the dominant plant group, angiosperms, which represent about 80 per cent of green plants, features about 300,000 species of flowering plants alone.

Life on the planet as we know it today is unthinkable without plants. They have had an effect on the important biogeochemical cycles of carbon, oxygen, nutrients, nitrogen, phosphorus, sulphur, rock and water, helping to shape not only the terrestrial sphere but also contributing to changes in the atmosphere and oceans. The roles that plants have played in these cycles have paved the way for other species to evolve and thrive. Their terrestrial colonization in the Palaeozoic period gradually resulted in lower concentrations of atmospheric carbon dioxide, providing conducive conditions that facilitated the emergence of land-based animals 350 million years ago, and eventually our ancestors in the *Homo* species 2.8 million years ago.

Plants have exhibited unique degrees of evolutionary ingenuity in their ability to respond to myriad ongoing challenges over the years, their resistance to threats and stresses, and their resilience in recovering from serious disturbances. They have adapted to the environments they have found themselves in by developing a diversity of morphological features suited to differing soil and climate conditions and availability of resources such as light and water. In turn, plants have played a role in shaping the environments they inhabit through their interactions with biotic and abiotic factors in an ongoing dynamic process of co-evolutionary influence.

The differences that set plants apart from animals are significant and they perform uniquely important functions. Along with algae, phytoplankton and cyanobacteria, plants have the autotrophic ability to produce their own food using light, water, carbon dioxide and other chemicals. Chlorophyl pigments in the leaves of plants primarily absorb the blue range (400–450 nanometres) and red range (600–700 nanometres) of wavelengths from the visible light spectrum emitted by the sun, converting the light energy into chemical energy as part of the process of photosynthesis. They combine carbon dioxide from the atmosphere, absorbed by pore-like stomata on the underside of the leaves, and water, drawn up from the soil by the plant's roots, to produce carbohydrate glucose. This is used to promote cellular growth in the various forms of biomass, such as stems, stalks, leaves, flowers, fruits and seeds. Glucose molecules also combine in more complex forms as starch and cellulose, the structural material used in plant cell walls.

Also during this process plants respire, expelling oxygen as a waste product, providing the air that we and other species can breathe. At night during respiration they expel carbon dioxide, with the net balance of carbon being stored in the soil. This carbon, along with that stored in the oceans, has helped to maintain a relative balance of CO_2 in the atmosphere, which makes it amenable enough for the existence of our species and the others we share the planet with.

As primary producers within ecosystems, plants are an important first stage in the nutrient cycle that provides a flow of energy through the food chain. When eaten by heterotrophic organisms such as herbivores, they provide energy and nutrients for growth and sustenance. This energy is then passed on up to the next trophic level of predators. When plants die, their matter enters processes of decomposition, recycling nutrients and energy carried out by decomposers, organisms both above and below the surface of the soil.

Being rooted to one place sets plants apart from animals, for whom being sessile is not a good strategy for survival, as it opens up easy opportunities for predation. But plants have turned this seeming disadvantage to their favour. Being static means that they are placed to receive a relatively steady supply of light, moisture and nutrients, rather than having to expend energy moving around to source further energy. In order to ward off threats, plants have developed a number of sophisticated biochemical defence systems. Poisons, psychedelics and aphrodisiacs are some of the compounds they use to deter unwanted herbivorous advances.

Being in one place requires a method of fixing to the spot, another function carried out by their roots, along with their role as conduits during photosynthesis. The fact that a large part of a plant is below ground is also partly a protection against being killed by predation, when their foliage is eaten by herbivores. Their underground architecture also forms symbiotic relationships with mycelium and bacteria, assisting in the beneficial transfer of resources between them.

Plants' cellular structure is also distinct from that of animals, with fewer and less complex organs that have allowed them not only to develop myriad different morphologies, but also to possess a high capacity to regenerate. Unlike most animals, for whom body parts are usually irreplaceable, plants display an extraordinary ability to revive tissues and organs lost or damaged in injury. This is obviously beneficial for themselves, but it also allows herbivores to eat them without necessarily killing their food source, creating a sustainable cycle of co-dependence.

Plants possess extremely varied reproductive strategies compared to animals, including both sexual and asexual reproduction, and sometimes both simultaneously. Sexual reproduction is often a co-evolved process reliant on other species to transport pollen and seeds, while asexual reproduction by clonal propagation asserts a distinct independence.

Plants are so intricately interconnected to other species and processes within ecosystems that they could be considered the glue that holds most of the terrestrial ones, and some of the aquatic ones, together. As a species *Homo sapiens* has co-evolved with plants and is bound to them in many ways. We are dependent on them for our most basic needs, from oxygen to the nutrients we need to eat in order to exist. They also provide humans with other essential material needs in the form of clothing, medicines and materials for building. But what does the future that we are heading towards mean for plants and our relationships with them? Will they just be part of the collateral damage we are causing or are they potentially our greatest ally in solving the problems we face?

PLANTS AND A CHANGING PLANET

Plants have evolved to fit their environmental conditions, something reflected not only in the forms they have developed but also in the physiological processes of their life cycles, such as photosynthesis, respiration, transpiration and growth. The relationships they have with their surroundings are a product of adaptations over long stretches of time, through periods in which the atmosphere has been hotter, colder, wetter and drier than it is now, with CO_2 concentrations that have been both higher and lower. The ways in which particular species have adapted to these varying conditions, as well as how quickly, have been crucial in determining their survival or extinction.

The current changes in patterns of temperature, precipitation and carbon concentration are notable for the rate at which they are occurring, raising questions as to exactly how plants are likely to respond. Aside from their direct reactions to each of these conditions, irregular and unpredictable weather events, such as storms, cyclones and fires, add to the challenges they face. Different species will respond in various ways according to their particular characteristics, some dealing with certain challenges better than others. It is not simply the singular changes that must be faced but also the complexity of them when they are acting simultaneously that pose additional problems.

The effects of the balance between temperature and moisture have an impact on diversity. Regions such as the tropics, with high temperatures and moisture levels, tend to be more diverse, while arid hot deserts are less so. A species may be able to tolerate a higher temperature within its ecological range, but a fluctuation in the availability of water, either more or less, may prove detrimental. Most obviously, extreme temperatures can lead to drought, causing greater competition between plants for moisture, dehydration stress, metabolic disruption and mechanical damage to cell membranes. Also, both drier and moister conditions bring with them a plethora of potential pathogens and pests suited to the new conditions, assailing plants unprepared to resist.

The appropriate availability of water is crucial for plants, yet predictions for precipitation are less consistent than those regarding temperature. More moisture in some areas and less in others will impact differently on plants' suitability for the location and growth patterns. Some areas are even likely to experience both of these extremes at different times of the year, posing more of a problem. Decreases in moisture will negatively impact on plant growth, but to what degree is dependent upon total rainfall during the growing season, as well as the intensity and magnitude of rainfall events. The amount of water absorbed by the soil will also affect the make-up of bacterial communities – responsible for decomposition and recycling nutrients to plants – within it.

Temperature and moisture are both drivers that determine plant persistence in a region and influence vegetation patterns geographically. Given the vicissitudes of these factors, plants are finding themselves living at the limits of the ecological ranges in which they can survive. Some are in the process of adapting in order to remain where they are, but others are trying to deal with it by migrating to new locations in order to find habitats that offer the optimal goldilocks criteria they are accustomed to. This has been resulting in a distinct drift of some species in latitudinal directions and also to higher altitudes with cooler air, to escape from hotter temperatures at lower levels. For many plants this is possible, but for alpine species near mountain peaks, it is an escalator to extinction, as there simply isn't anywhere higher for them to go.

Given the central role of CO_2 in photosynthesis and the abundance of it in the atmosphere, the changing climate is having noticeable effects on plants' distribution and phenology, which have serious implications for ecological processes and other species. As CO_2 is integral to energy resources for plants as well as their nutritional metabolism for growth, increased concentrations in the atmosphere available for photosynthesis have profound effects on growth and physiology, acting as a boost to their rate of growth and increasing biomass, both above ground and to a greater extent below ground. It also results in less water use during the process, but if conditions are drier the ability to utilize more CO_2 may be thwarted. The need for more nutrient uptake from the soil may have the same effect if it is not readily available. An additional 'swings and roundabouts' scenario is that while the ability of plants to utilize the additional CO_2 is a good sign, it can easily be negated by drought or flooding reducing the number of plants actually carrying out the process.

Most plants are scientifically classified as C3, referring to the particular chemical process that extracts and processes carbon dioxide during photosynthesis, which defines their efficiency and water use limitations. But about 3 per cent of plant species are C4 and have evolved an adapted process that improves their efficiency so that they can better endure high light and temperature environments. Higher temperatures would favour the latter, potentially leading to competition between these groups.

Plant phenology is the timing of plant life-cycle events, such as leaf bud burst, flowering and fruiting. Signs of phenological change have been noticeable in the deviation of plants from their typical growth patterns associated with seasonal change, something that has been observed to be occurring in parallel to increasing temperatures for decades. In spring, the timing of leaf appearance and flowering has been observed to be advancing each decade, according to records collected since the 1980s, by 4.2 days in Europe, 0.9 days in the United States and 5.5 days in China.[1] A comprehensive survey published in 2022 of the first flowering dates in the UK, using phenological observations gathered since 1753, showed that flowers have been appearing almost a month earlier in the period from the mid-1980s to 2019 compared to the preceding period; this correlates with increased temperatures over the same time span. Also noted was that many plants in the UK are flowering almost one week earlier in the south than in the north and at lower elevations, and that the largest shift has occurred in herbaceous plants, which generally have faster turnover rates and potentially higher levels of genetic adaptation than shrubs and trees.[2] In autumn changes in leaf colour and fall have been delayed by varying amounts. Distinct patterns – such as annual plants flowering before perennials and those pollinated by insects before those wind pollinated – have been emerging. Research has also revealed that phenological disruption is associated with loss of biodiversity caused by climate changes.[3]

While the effects of a changing climate are varied with different individual species, a major concern is the impact on the central role that plants perform within interconnected ecosystems, influencing either directly or indirectly the other biological levels within them. The potentially cascading events caused by disruption to plant functions, diversity or abundance raise major concerns about their ability to function as effectively in the future.

A stable ecosystem exhibits little variability and is resistant to change from external environmental forces. Its resilience is indicated by the amount of time it takes to recover to its original state from a disturbance or perturbation. Ecosystems are more stable the more diverse they are, as a variety of responses to environmental challenges means that some species can deal with them, compensating for those that cannot, thereby ensuring an overall functionality. Increased diversity in species means that individual species can fluctuate in their abundance, some increasing while others decrease, without undermining the stability of the ecosystem. If particular species of plants are unable to survive the challenges they face, then the relationships and interactions within the ecosystems can become seriously compromised. Observed and predicted changes in plant distribution and phenology have major implications for ecosystem productivity, community structure and levels of biodiversity, while changes in flowering time and leaf development are likely to alter patterns of interaction between plants and their pollinators and herbivores. Alterations to concentrations of chemicals, such as nitrogen in plant tissues, will also affect herbivores' protein supplies and the amount of plant matter they consume.

Increased coverage of grasses and deciduous shrubs, alongside decreased coverage of ferns, mosses and lichens, as well as changes to canopy heights and density, alter community architecture and the physical forms of the species, which in turn alter their usage as habitats by other creatures. More frequent and intense drought periods, resulting in a reduction of plant species, can disrupt the phytosociology of entire plant communities over time, decreasing ecosystem diversity and making its functioning less efficient. Less efficient functioning sets off positive feedback loops that then have an ongoing series of degrading effects on the environment, in ways that are as yet unknown.

PLANTS
AS
PANACEA

THE UNSEEN GREEN

Despite the important functions that plants perform, they are all too often misunderstood and underappreciated. Many people in urban and industrialized societies don't actually notice plants, except as a vague green backdrop to their everyday lives, even though their prevalence across the surface of the planet is unmissable.

This lack of awareness was termed 'plant blindness' by botanists James Wandersee and Elizabeth Schussler in 1999; they defined it as '(a) the inability to see or notice the plants in one's environment; (b) the inability to recognize the importance of plants in the biosphere and in human affairs; (c) the inability to appreciate the aesthetic and unique biological features of the lifeforms that belong to the Plant Kingdom; and (d) the misguided anthropocentric ranking of plants as inferior to animals and thus as unworthy of consideration'.[1] This means that people, on a daily basis, fail to see plants around them. As a consequence, they also fail to understand their ecological and practical importance, and lack a knowledge of their needs and relationships to specific geographic and climatic environments. Part of the reason is that entrenched zoocentric teaching privileges animals over plants. There is a tendency to relate to animals for beneficial reasons as well as for those of safety, but also animals are given greater attention than plants because they are more similar to humans, and this bias is reflected in notions of species hierarchy in education and culture.

Plants' lowly status can also be attributed to a number of other factors. For humans, they are generally non-threatening and can usually be ignored without serious consequences. They do not have faces and, being rooted to one place, are unable to move around. Being perceived as an undefined, amorphous mass that blends into an indistinct backdrop for other objects or activities means that no differentiation is made between specific types of plants such as trees, shrubs, perennials and grasses. This is particularly evident when flowering plants are without their flowers, which add distinct visual accents of colour. The perception of plants as an undifferentiated form can create a sense of overfamiliarity, diminishing the attention afforded them, and leading to a lack of understanding of how these different types of plants create specific landscapes and the variety of distinct roles that they play within them.

Awareness of plants is not necessarily straightforward and universal, but rather reflects a spectrum of different degrees of acknowledgement determined by factors such as race, sex and cultural attitudes. The more importance a culture ascribes to plants, and the more people within it who work directly with plants or plant products, the more awareness there will be. In indigenous cultures that still maintain a direct relationship with the land, an acute awareness of plants, their abilities and benefits is a traditional form of embodied knowledge in everyday activities. But in societies where people's lives are increasingly at a remove from the rest of nature, the importance of plants is less evident.

The term 'plant blindness' has come under criticism for its ableist connotations, and suggestions have been made for a more appropriate name, such as lack of 'flora appreciation'.[2] The comments are valid and certainly need to be addressed, but the importance of the gist of the idea is undiminished, in that there is a desperate need to revise our attitude towards the flora around us.

Another thing that distances people from plants is the severing of their traditional relationships with them in a process known as biocultural homogenization, which implies a loss of both awareness of and willingness to conserve local nature as a result of the 'extinction of experience' or the lack of face-to-face encounters with local biodiversity.[3] It marks the global progression of sameness that denudes landscapes of their sense of distinct uniqueness, cutting the ties between them and the people who inhabit them. It is largely due to ever-expanding urbanization, as well as events such as deforestation in the Amazon to make way for monocultural fields of introduced crops growing feedstock for farmed animals in far-flung parts of the temperate world. It is a form of environmental colonialism that creates a loss of awareness of the importance of local biodiversity and a lack of determination to conserve natural features, as well as local traditions and occupations that are associated with them. The process goes hand in hand with advancing capitalism, the rise of introduced languages, cosmopolitan cultures, and the global circulation of certain types of plants.

The fatigue associated with global homogenization feeds into the lack of awareness of plants. If people don't see or think about plants generally, then it is difficult to convey the important roles they play and our dependence upon them. Or if plants are devalued and demeaned as obstacles or nuisances, then this makes it easier for biodiverse areas such as rainforests to be destroyed without due concern. Lack of appreciation of plants creates a barrier to communicating the need to confront environmental problems at every level and in every location.

The diminishing presence of plants in language is also a problem. Ironically, the system that has helped most to categorize and unify plant knowledge globally has also played a role in disassociating people from plants. Using binomial Latin, the system devised by Swedish botanist Carl Linnaeus to identify plants has made cross-cultural and linguistic communication about plants a lot easier. However, the process of imposing Latin names on plants and stripping them of their local common names – ones that link plants, people and place in biocultural relationships through time – was a decidedly Eurocentric and colonial enterprise. Its legacy today can be seen in hierarchical social and class-based divides, between those who know the scientific names and those who use the common names, as well as in the diminishing importance of those original names and their connotations. While science has now debunked many of the medicinal associations of certain plants held by herbalists, which are often reflected in their original names, these names are nonetheless an important part of history. Increasingly they are being lost as dictionaries drop them from their listings in an act of self-fulfilling prophecy, because they are not being used enough to merit inclusion.[4]

Addressing the issue of plant appreciation is a serious task to be undertaken. Wandersee and Schussler suggest that it should be carried out through the various forms and institutions of education and mentoring. But perhaps the biggest lesson about our relation to plants and the rest of nature has been forced upon us whether we like it or not.

HEALTH AND WELL . . . BEING

The coronavirus pandemic that began in 2020, with its lockdowns and isolation, brought with it a dawning widespread realization of the benefits of being outdoors and, more specifically, of being in places with plants. The opportunity to get outside to walk, exercise and engage with some form of semi-natural environment proved for many people key to being able to maintain a balance of sanity and connection to the world. Green spaces provided a sense of stability and solace unavailable in the social realm. It made apparent the connection between healthy humans and healthy resistant ecosystems. As researcher Jake Robinson and colleagues stated in a paper, 'There is growing recognition that all forms of life are interconnected, ecologically and evolutionarily. These tangled connections also traverse the boundaries of the sociosphere – the complex realm of dynamic human-centric structures and interactions that weave their way in and out of our social lives and cultural identities. Indeed, it is the interconnectedness of societal health with environmental stability and resilience that are integral to the concept of planetary health.'[1]

Yet this bigger picture regarding the feel-good factor that connections with plants bring was nothing new, and had actually been slowly gaining widespread acceptance for some time. Drawing upon an idea initially floated by psychoanalyst Erich Fromm, the entomologist E. O. Wilson used the term biophilia in 1984 to describe our species' innate attraction to nature and all the life forms and processes it embodies. He recognized that it has been hardwired into our DNA through evolution, and that we unconsciously understand our interconnectedness in the greater web of life on the planet and seek out meaningful relationships with it, despite the fact that through millennia of civilization we have come to forget this, or – perhaps – that we override our tendencies to acknowledge it in our constant quest for self-improvement and social progress.

A body of scientific evidence has been growing ever since, developing in fields such as neuroscience, microbiology and ecopsychology, that reveals how making meaningful connections with nature is essential for our physical and mental well-being. Journalist Richard Louv coined the term 'nature-deficit disorder' in his 2005 book *Last Child in the Woods* to highlight the impoverishment suffered by decreasing engagement with nature. A multitude of scientific research followed with experiments charting such benefits as patient progress in hospitals, based upon the access to views of trees outside their windows, and recent writers such as Florence Williams, Lucy Jones and Sue Stuart-Smith have charted the research for a non-academic audience, examining our ongoing disconnection from nature caused by the demands and stresses of life in the modern world.

The physical, mental and emotional benefits of engaging with green spaces and plants revealed by the growing body of research are manifold. These include stress and anxiety reduction, attention deficit recovery, decreased depression, greater happiness and life satisfaction, enhanced memory retention, mitigation of PTSD, increased creativity, reduced effects of dementia and improved self-esteem.

What is surprising about these revelations is that they are not surprising at all. As a species that has its origin in outdoor spaces, they are undeniably part of our habitat and we have inextricable relationships with them – as biophilia suggests – yet we have become cloistered in sterile indoor spaces that aim to keep the rest of nature at bay. Obviously, the safety and protection that these spaces provide can be a good thing, but the more that urbanization has increased, the more divorced we have become from natural environments unaffected by human activities. This tendency has been accelerated and exaggerated in societies that evolutionary biologist Jonathan Schultz has described with the acronym WEIRD (Western, Educated, Industrialized, Rich, Democratic). Tracing the anthropological kinship relationships that produced Northern European formal institutions, laws and markets, Schultz points out that the characteristic traits within these societies tend to the individualistic, independent and competitive rather than the supportive and communal. The manifestation of nature-deficit in poor physical and mental health can be read as a product of the pursuit of progress and profit. According to one study published in 2005, 'Whilst modern "westernization" has doubled our life expectancy, it has also created disparities between ancient and present ways of living that may have paved the way for the emergence of new serious diseases.'[2] Tony McMichael, an epidemiologist from the Australian National University, explained in 2001: 'Modern "Westernisation" has created disparities between ancient and present ways of living that augment various serious diseases, yet it has also doubled our life expectancy.'[3] However, he noted that: 'As more people survive to older age, and as patterns of living, consuming and environmental exposures change, so non-communicable diseases such as coronary heart disease, diabetes and cancer have come to dominate.'[4]

In terms of our species' history, urban living is a relatively new phenomenon, and the consequences of adapting to it successfully may well underlie many of the physical and mental traumas of the current age. All of these factors, aligned with cultural ideologies that separate humans from the rest of nature, are a recipe for individual and social discontent. Obviously, counteracting this by making connections with the rest of nature is beneficial and to be encouraged, but often even this can result in a positive feedback loop. An example is how adopting a concerned attitude towards the planet and considering the damage done by the climate and extinction crises can cause further psychological distress. A word has even been coined by philosopher Glen Albrecht for this type of existential distress: 'solastalgia'. It describes 'the pain experienced when there is recognition that the place where one resides and that one loves is under immediate assault (physical desolation). It is manifest in an attack on one's sense of place, in the erosion of the sense of belonging (identity) to a particular place and a feeling of distress (psychological desolation) about its transformation.'[5]

A lot of recent nature writing has been keen to explore the phenomenon of what has become dubbed, after the title of Richard Mabey's 2005 book, the 'nature cure'. It may seem self-evident that those with a vested interest would extol the virtues of nature. Yet Mabey himself is ambivalent about the idea and has always been keenly aware that it is something of a double-edged sword. For while the benefits of nature may seem obvious, there is an irony in expecting nature to provide a panacea for our problems, given that it is precisely our maltreatment of nature that is the cause of them. We have an unhealthy attitude towards nature and yet expect it to heal us. The coronavirus situation can also be seen as a manifestation of this contradiction, in the sense that the edicts to get outside to soothe our states of physiological distress caused by this devastating and disconcerting pandemic are necessary because of our maltreatment of other more-than-human species in the first instance.

Even such august organizations as the Royal Horticultural Society advise that plants are beneficial for reducing stress and increasing productivity. But what is the endgame of productivity? Possibly to induce yet more stress that then needs to be relieved by plants, and so on in a vicious circle. The problem is not so much a physical one but rather a cultural and ideological one, based upon our idea of ourselves as a species and our place on the planet.

THE NATURE DISCONNECT

Despite the fact that environmental awareness has been growing over the past fifty years, the majority of humans still continue down a destructive path, pursuing an asset-stripping mentality that seeks to extract and plunder not only for the benefit of our species alone but, even more inequitably, for that of specific groups and individuals within it. It is worth asking why we allow it to continue, if we realize that our behaviour is having negative impacts on ourselves and the world around us.

Ingrained habits, prejudices, the motivation of individual, corporate and national progress and profit, greed, and the endless desire for increasing hits of novelty are some of the reasons. At the heart of it is an unhealthy attitude towards the world around us, and more specifically the world that isn't human, in all its living and inanimate guises. These are all wrapped up in the idea of 'nature', a socially constructed concept that cultural theorist Raymond Williams referred to as the most complex term in the English language.[1] It cannot easily be captured in a simple thought but instead is an amalgam of ideas, often contradictory, loosely tied together. The thing that is usually conjured by the word is 'that which is not human'. It denotes something 'other', against which humans define themselves, not only as separate from everything else but also as superior.

There are many ways in which we are similar to other animals and even plants: we share much of the same DNA with a lot of them. In terms of differences, the factor that really sets us apart is self-consciousness. This reflexive trait appears to be unique to our species, a biological product of our large brains that has been shaped through evolution by a series of adaptations in order to be a useful tool for navigating the challenges of the world around us. It provides us with the ability not only to recall the past but also to imagine possible futures, create narratives to provide explanations or for amusement, and to have a sense of selfhood and identity, with all the personal and social responsibilities that implies.

Self-consciousness has been used as a reason to aggrandize our position within the world, setting ourselves up as better than other species by deliberately prioritizing considered action above instinctual. But there is nothing that is intrinsically important about self-consciousness, as opposed to the other techniques employed by more-than-human beings in order to survive. From an evolutionary perspective, the endgame is the continued existence of a species, which many others achieve at least as effectively as humans. The teleological explanation that the history of evolution has been a long trajectory to reach this ultimate state of human dominance is a reverse-engineered anthropocentric conceit used to justify human disassociation from the rest of nature and uphold the exploitative relationship we have with it.

A consequence of self-consciousness has been the ability to communicate and socially organize at a scale that has seen humans spread across the globe. The inquisitive behaviour and reflexive ability to evaluate things and situations which self-consciousness provides has led to discoveries and inventions that have been useful in solving problems and facilitating human expansion in numbers and geographically. As a tool for guiding our survival it could be seen to have been extremely successful, but given the fact that through a long series of historical events it has led to the current situation, it must also be recognized that its use can be just as destructive as it is productive.

Using these cognitive faculties, the development of basic technologies began a process of distancing and managing the environment. The ability to control fire, which was in daily use 300,000 years ago by our predecessors *Homo erectus* and *Homo neanderthalensis*, marked an evolutionary paradigm shift. This technology provided heat and light, allowing habitation of areas with colder climates, and made it possible to cook food, reducing disease and increasing nutrients, resulting in a growth in brain size and consequently enhanced intellectual abilities. What is more, fire not only kept predators at bay but was also used to burn bush and scrub in order to pursue prey. Clearance of areas of forest by fire also provided land where desirable plants were allowed to flourish without undue competition, in a form of proto-farming.

Later, when groups of *Homo sapiens* renounced their seasonally based nomadic hunter-gatherer lifestyles and settled down in static conurbations around 10,000 to 12,000 years ago, they began domesticating plants to grow as food sources. Agriculture provided a new system for managing food supplies and food security, with more dependable outcomes than the unpredictability of hunting and gathering. This offered obvious benefits in terms of an increased regular source of nutrients, improved health and survival rates, and population growth. Agricultural areas became important social centres, many eventually growing into towns and then cities, developing forms of exchange and interaction that paved the way for the societies many inhabit today – cementing the close association of the act of cultivation with the idea of civilization. But this signalled another significant step in the transition from simply being one animal among many others, living in dynamic co-dependent relationships in complex ecologies, to being the dominant animal that has gradually shaped the landscape, air and oceans by practically enacting the binary separation between humans and the rest of nature.

This freedom for a large number of people from actively participating in providing the means of subsistence from the environment allowed time to develop other interests in the growing sphere of culture, including intellectually grappling with understanding our place on the planet. Foundational religious texts, philosophical rumination on the meaning of existence, and scientific investigation into the workings of the world reinforced this notion of nature as distinct from humans, placing them upon a pedestal, at the pinnacle of the pyramid of life forms, justifying their dominance over all those on the levels below them.

Prior to the Judeo-Christian period, belief systems with multiple gods and spirits dominated social life, until the new religious doctrine dispelled them or synthesized them into a singular supreme being that exercised authority. The Old Testament Book of Genesis asserted human dominion over nature and established a clear trend of anthropocentric superiority over the world: 'And God said, Let us make man in our image, after our likeness: and let them have dominion over the fish of the sea, and over the fowl of the air, and over the cattle, and over all the earth, and over every creeping thing that creepeth upon the earth' (1:26). The distinction between 'man', formed in God's image, and the rest of creation, which has no soul or reason, provided a privileging of the former as superior and demeaned the latter as inferior, justifying the use of the natural world as a means to human ends.

The Western philosophical tradition also has a lot to answer for in developing the idea of nature as inferior. In the world of classical Greece, Aristotle considered the world to be a large mass of matter subjected to continuous change and existing for the benefit of humans. His biological hierarchy elevated humans above both animals and plants. It placed plants at the lowest level, with vegetative souls geared solely towards reproduction and growth; then came animals, with sensitive souls enabled by mobility and sensation, both of which are superseded by humans, who have rational souls capable of thought and reflection. Aristotle's ideas were related to his taxonomical system of categories dividing all the life forms in the world into groups based upon various characteristics. The influence of his system later fed into other taxonomies, including the work of Swedish botanist Linnaeus in the mid-eighteenth century. Linnaeus created his own system (as well as the widely recognized means of species identification by binomial nomenclature). While his and other later systems of classification were intended to be neutral, they were nonetheless often used as justifications for the privileging of one species over another. (The term 'speciesism' to describe such prejudice was coined in 1970 by British psychologist Richard D. Ryder.)[2]

The scientific revolution in the seventeenth century ushered in a new way of thinking about nature, which viewed it as mechanized and to be dominated. English philosopher Francis Bacon's concept of knowledge was firmly rooted in the Christian tradition, and sought to regain the prelapsarian human dominion over nature based on reason, as he made clear in *Novum Organon*: 'Let the human race recover that right over nature which belongs to it by divine bequest.' Investigating the world meant that its workings could be exposed and explained as a series of simplified interactions, which could provide insight into how humans could best utilize it for their own purposes. His ideas provided the premise of the scientific tradition, in which observation and investigation of the world meant that if things were known they could be controlled and used for the benefit of humans.

A few years later, in France, René Descartes suggested a theory of dualism that framed the world as essentially split between the realm of the mind and that of inert matter. He proposed that, because of self-consciousness, humans have the ability to consider and therefore confirm their own existence, as he explained in *Discourse on the Method and Principles of Philosophy* (1637) with his famous tautological maxim: 'I think, therefore I am.' Given that it was only self-consciousness that could legitimate itself, he established a dualism between mind and matter in which self-consciousness ordained humans as thinking beings and therefore as separate from the rest of nature, and consequently superior to it. Animals were deemed mere mindless machines, unworthy of any rights or moral consideration.

This was not only one of the most enduring endorsements of the human/nature binary, it also laid the foundations for the idea of the sovereignty of the individual and its enshrinement in legal and political rights. The association of the individual with ownership and property transformed the notion of nature as something available to all to one tied up with personal possession. English philosopher John Locke suggested that the natural world was for the taking, assuming one had laboured to get it: 'Though the earth, and all inferior creatures be common to all men . . . Whatsoever then he removes out of the state that nature hath provided and left it in, he hath mixed his labour with, and joined to it something that is his own, and thereby makes it his property.'[3]

The combination of the ideological division between humans and nature, and individual rights to ownership, aided the development of capitalism and the industries that fuelled it by extracting from the rest of nature as if it were an unlimited resource. These industrial acts implemented the culture/nature barrier on a huge scale, helping to normalize the acceptability of natural resources being exploitable as simply part of the process of becoming a modern civilized society. Charting the long progress of humans from their early origins, philosopher Carolyn Merchant explains:

> The change in controlling imagery was directly related to changes in human attitudes and behaviour toward the earth. Whereas the nurturing earth image can be viewed as a cultural constraint restricting the types of socially and morally sanctioned human action allowable with respect to the earth, the new images of mastery and domination functioned as cultural sanctions for the denudation of nature. Society needed these images as it continued the processes of commercialism and industrialization, which depended on activities directly altering the earth – mining, drainage, deforestation.[4]

The resulting spiralling effects of production and consumption have pushed us to the planetary boundaries that we are confronting today.

Reinforcing the binary split between nature and culture has been an ongoing process in which the rationalistic Western worldview has separated active, reflective subjects from passive, instinctual objects by drawing parallels with other socially constructed dichotomies. For Australian philosopher Val Plumwood: 'The set of interrelated and mutually reinforcing dualisms which permeate western culture forms a fault-line which runs through its entire conceptual system.'[5] These oppositions reflect a standpoint of domination and mastery associated with a patriarchal colonial mindset, which privileges one side of the equation over the other in an attempt to maintain a social stability favoured by those who benefit most from them. Other dualisms include masculine/feminine, mind/body, reason/emotion, civilized/primitive. In these dualisms, 'what is taken to be characteristically and authentically human, or of human virtue, is defined against or in opposition to what is taken to be natural, nature, or the physical or biological realm . . . the ideal for which humans strive is not to be found in what is shared with the natural and animal.'[6] The layering of these dualisms accords them a far greater potency than they would have alone and allows them to have a wider influence. They denigrate not simply nature but also anything associated with it, in instances ranging from the subordination of women to the justification of colonialism and slavery. As anthropologist Deborah Bird Rose explains, 'The well-worn dualisms of Western thought have played a crucial role in the violent transformation of peoples and ecosystems. They continue to obstruct our ability to achieve both social justice and environmental justice.'[7]

Intersectional approaches to justice have highlighted the ways in which the confluence of various power structures use assorted social constructs of nature to reinforce hierarchical assumptions about ethnicity and privilege. Ecofeminism has shown how the conflation of sex and gender with ideas of nature undermines the positions of both women and the natural world under patriarchy, while the nascent discipline of Queer Ecology has problematized the connotations embodied in the binary split between 'natural' and 'unnatural', for the ways in which it reiterates hegemonic heteronormative discourse.

But despite these ideologies of disconnection, nature is a package deal and we are all in it, together with everything else, no matter how hard we try to delude ourselves that we are not. We are and always have been very much part of nature, and everything we do is natural, despite how complex and removed from primary sources and processes some of it is. As carbon-based, energy-fuelled multicellular organisms, we are like many other life forms carrying out similar basic tasks. As biologist Lynn Margulis makes clear, 'We are part of an intricate network that comes from the original bacterial takeover of the Earth. Our powers do not belong specifically to us but to all life.'[8] We adapt to our habitats and shape them to our needs in order to satisfy daily requirements for food, shelter and security. We are products of evolution and have been formed through numerous encounters with the world over a long period of time, just as other species have been. Evolution presumes no linear trajectory towards an endgame and has no hierarchies based on moral judgements. As environmental philosopher J. Baird Callicott explains, 'Bluntly put, we are animals ourselves, large omnivorous primates, very precocious to be sure, but just big monkeys, nevertheless. We are therefore a part of nature, not set apart from it. Hence, human works are no less natural than those of termites or elephants. Chicago is no less a phenomenon of nature than is the Great Barrier Reef.'[9]

Recent research has even gone so far as to dismiss the notion that we are fully unique as biological humans and distinct individuals. Our bodies contain multitudes of microbes that blur the boundaries between ourselves and other species within our own bodies. The mixture of microbes on and within the body, known as the microbiome, plays an important role in human physiology and disease, so much so that they ensure our health and existence. Approximately 43 per cent of cells that form a human body are uniquely human, while the remaining 57 per cent are microbial, a combination of bacteria, viruses, archaea and micro-eukaryotes. The balance is further tipped in favour of these mini-organisms by their sheer numbers. Microbial genes outnumber human ones by a magnitude of between 150 and 1,000 times. On the palm of one hand reside 150 different species, while the oral cavity is home to 700 and the gut to 1,000. As Jake Robinson, Jacob Mills and Martin Breed suggest, 'With a moment of reflection, this can lead to a medley of existential questions such as – what does it mean to be human?'[10]

MANAGING THE ENVIRONMENT

The disconnect between humans and the rest of nature has been reflected in the ongoing uneasy tension between ecology and economics, two quite different disciplines that share the etymological root 'eco', from the Greek word 'oikos', meaning 'house'. Economy is derived from the Greek word 'oikonomos', meaning the management of household affairs, especially expenses and the efficient use of resources. Ecology comes from the German word 'ökologie', coined in 1866 by the biologist Ernst Haeckel, referring to the study of homes and habitats, and is concerned with the interrelationship of organisms and their environment.

While ecology is very much focused on physical places and the life forms of which they consist, economy has become an abstracted self-referential notion, with no defined place in the real world yet nonetheless affecting it in many ways. This rift between the two is at the heart of the current environmental crises, and their incommensurability has been the stumbling block on which significant progress to address matters has constantly faltered. Capitalism's inexorable drive for growth has resulted in the economy constantly trumping ecology, in the interests of those profiting from the continuation of the status quo.

During the late twentieth century, numerous attempts were made to shore up the balance between anthropocentric endeavours and the environment, as the source of provision for our planetary home needs, by unpicking the closely threaded weave involving the extraction of natural resources, their transformation into commodities, and the labour relationships involved in these processes. The equation of global population growth plus increasing consumption in Western economies and the resulting environmental degradation increasingly became something of a deficit balance.

Resource depletion and the ability to provide for an ever-expanding global population were addressed in Garrett Hardin's 'Tragedy of the Commons', published in the journal *Science* in 1968. The privatization and exploitation of open-access and unregulated resources, such as the atmosphere, oceans and rivers, exemplified the enshrinement of the colonial extractive mindset into mainstream economics. In the same year, Paul Ehrlich echoed these concerns in his book *The Population Bomb*, while economist Herman Daly's article 'On economics as a life science' proposed that biology and economics were both firmly rooted in the same processes of energy flows of inputs and outputs necessary to sustain life. The relationships between human activities and the natural world were intimately interconnected, yet the failure to account for the externalities of damage to the environment posed a major threat to the future: 'It is only through the agency of air, soil, and water that plant life is able to capture the solar energy upon which the whole hierarchy of life (and value) depends. Should not these elements receive the same care we bestow upon our other machines? And is not any theory of value that leaves them out rather like a theory of icebergs that fails to consider the submerged 90 percent?'[1]

Also in 1968, an organization named the Club of Rome was convened by Italian industrialist Aurelio Peccei and Scottish scientist Alexander King. Comprising a range of academics from various disciplines, it was set to undertake the arduous task of addressing the various ills blighting humanity. Dismissive of the failures resulting from taking individual approaches to matters of global concern, such as environmental destruction, poverty and health, they instead sought a more joined-up approach. In 1972 they published a report, *The Limits to Growth*, an overarching analysis of global economics within the context of planetary resources, compiled by an international team at the MIT Sloan School of Management, headed by Donella H. Meadows, Dennis L. Meadows, Jørgen Randers and William W. Behrens III. Addressing the complexities and uncertainties surrounding the issue, the report used computer simulations to look at different scenarios and to calculate how finite resources could supply the needs of a rising global population with expectations of economic growth, without overshooting planetary boundaries and causing a collapse of natural and social systems.

The now all too familiar term sustainability received an airing in the report as a call to caution: 'We are searching for a model output that represents a world system that is: 1. sustainable without sudden and uncontrollable collapse; and 2. capable of satisfying the basic material requirements of all of its people.'[2] The report's thrust was that unless serious system change occurred, ecological, economic and social decline were predicted within a century. Sustainability was a restriction to keep things within the boundaries of naturally occurring resources. Despite being intended as a positive wake-up call, the report received a cold reception for its perceived negativity and its emphasis on the need to address the problem in a systemic manner.

In 1973, the book *Small is Beautiful*, by the economist Ernst Friedrich Schumacher, echoed the concern that considering natural resources as expendable commodities was an unrealistic basis on which to premise economics. Recognizing that the effects of environmental degradation called the classical economics idea of 'free' natural resources as permanent and indestructible into question, he advocated the reappraisal of 'the existence of "goods" which never appear on the market, because they cannot be, or have not been, privately appropriated, but are nonetheless an essential precondition of all human activity, such as air, water, the soil, and in fact the whole framework of living nature'.[3]

In 1984, the United Nations General Assembly commissioned an investigation to examine the critical relationship between the environment and development, and to propose international means of co-operation across institutional bodies. This was assigned to the World Commission on Environment and Development and chaired by the former Norwegian Prime Minister Gro Harlem Brundtland. Their report, *Our Common Future*, was delivered in 1987. The idea of sustainability appeared here also, not as a form of restraint but instead the opposite, in a new strategy that entertained 'the possibility for a new era of economic growth, one based on policies that sustain and expand the environmental resource base'.[4] The report attempted to address different national levels of development to balance economics, recognizing that social issues of poverty reduction, gender equity and wealth redistribution were crucial to strategies for environmental conservation. It broadened the notion of sustainability as the link between the environment and the economic and social lives of humans, encapsulated in the phrase 'development that meets the needs of the present without compromising the ability of future generations to meet their own needs'.[5]

ECOSYSTEM SERVICES

A decade after the Brundtland report, new ideas emerged that sought to bridge the gap between the environment and economics. One was the idea of 'ecosystem services', viewing natural processes in terms of immediate benefits to human well-being and economy, which had begun to be mooted in the 1970s and then solidified at the end of the 1990s. The 1997 book *Nature's Services: Societal Dependence on Natural Ecosystems*, edited by environmental scientist Gretchen Daily, laid the foundations for the process of ascribing economic value to natural processes. In it she defined ecosystem services as

> the conditions and processes through which natural ecosystems, and the species that make them up, sustain and fulfil human life. They maintain biodiversity and the production of ecosystem goods, such as seafood, forage, timber, biomass fuels, natural fibre, and many pharmaceuticals and industrial products. The harvest and trade of these goods represent an important and familiar part of the human economy. In addition to the production of goods, ecosystem services are the actual life-support functions, such as cleansing, recycling, and renewal, and they confer many intangible aesthetic and cultural benefits as well.[1]

The same year, one of the book's contributors, ecological economist Robert Constanza, published 'The value of the world's ecosystem services and natural capital' in the journal *Nature*; in this paper he analysed seventeen ecosystem services in sixteen biomes, to arrive at a value for the entire biosphere of US$33 trillion per year.[2] Intending this as a way of highlighting the importance of these processes and encouraging their sensible use, the author believed that attributing financial value would act as a way of protecting them.

The concept became institutionalized in the international sphere of policy making in 2005 through the United Nations publication *The Millennium Ecosystem Assessment*. The report defined ecosystem services as the benefits people obtain from healthy ecosystems, breaking these down into four categories of services: supporting, provisioning, regulating and cultural. Supporting services take place through primary production, nutrient cycling and soil formation. Provisioning includes the supply of food, materials, medicinal resources and energy. Regulating means managing climate, purifying water and air, decomposition and pollination. Cultural includes spiritual and educational and aesthetic relationships with ideas of nature, as well as physical recreational engagements with the environment.

The related idea of 'natural capital' had first appeared in *Small is Beautiful*, in which Schumacher used it as a cautionary phrase to refer to limited planetary resources, but it acquired a more traditional economic meaning, describing the world's stocks of natural 'assets', including geology, soil, air, water and all living things, which form the foundation for the processes of ecosystem services. As a result, natural capital accounting was developed as a framework to measure and report on the stocks and flows of natural capital which a global financial sector has growing around it. While the original idea of ecosystem services – as a financial incentive for making economic decisions by taking into account the previously neglected externalities provided by natural processes – may have been well intentioned by its pioneers, it has proved problematic by reinforcing the importance of economy over ecology. Unashamedly anthropocentric, it reinforces the separation of humans from the rest of nature and their assumed superiority. The values within the system are considered solely as unidirectional – from ecosystems to the benefit of humans – and one of its leading proponents, the Intergovernmental Science-Policy Platform on Biodiversity and Ecosystem Services, has even suggested rebranding it as 'Nature's Contributions to People'. These benefits are accounted for only in monetary terms rather than other metrics, such as well-being, quality of life and health outcomes. This form of commodification reifies natural processes as mechanistic and reiterates an exploitative attitude towards them. While this was the opposite of the initial intention, the use of natural processes in financial markets helps to strengthen the neo-liberal system that degraded and devalued them in the first place, an irony that seems to be lost on many natural capitalists. It also sets up a hierarchy of ecosystems, placing higher value on the ones that benefit humans. Some ecosystems, while diverse and functioning, are ignored as they offer little in terms of perceived utility, neglecting any interrelated effects they may have within a bigger networked perspective.

Despite some progress made using the system, it has yet to provide a satisfactory resolution to the disjunction between ecology and economy. There can be benefits to using a cost-benefit analysis to put natural processes into a perspective that communicates some sense of their value, but these are primarily social, and the potential to address them in other ways exists without monetizing natural processes. Issues of local equity and security are related primarily to global wealth distribution rather than to semi-arbitrary attributions of worth to certain natural processes. The problem of who has the power to create these values is a matter of concern, given the historical and current imbalance in power and wealth between the Global North and Global South.

The more recent idea of 'nature-based solutions' has become the latest catchphrase to popularize the notion of ecosystem services further, using them as means to solve the problems of a changing climate and biodiversity loss. The definition proposed by the European Commission states that these solutions, 'inspired and supported by nature, which are cost-effective, simultaneously provide environmental, social, and economic benefits and help build resilience. Such solutions bring more, and more diverse, nature and natural features and processes into cities, landscapes, and seascapes, through locally adapted, resource-efficient, and systemic interventions.'[3] In 2020, the EC definition was updated to emphasize that 'nature-based solutions must benefit biodiversity and support the delivery of a range of ecosystem services.'[4]

While the general idea may be meant with the best intentions, its political uptake has, however, resulted in a fair amount of greenwash, an example of which can be seen in the penchant for planting trees, something that has reached fever pitch during election campaigning in the United States and Britain. While the carbon-capturing potential of trees is crucial, the ever-increasing quantities promised by rivals hoping to outbid each other at the polls lack not only genuine commitment but also an ecological understanding of how and where to actually grow trees. There is a crucial need for more tree planting, but its real value lies in the creation of functioning ecosystems, appropriately sited. The proposed benefits are often overstated, and as a political strategy it can be read as an offsetting attempt to deflect from actually addressing the real problem of continued carbon emissions. Criticism of many of these urban tree strategies suggests that they fail 'to take an integrated approach to socioecological sustainability that addresses the complexities of competing and compounding interactions among services, disservices, management costs, and differing perceptions among and within stakeholder groups'.[5]

ENVIRONMENTAL PRACTICE

The uneasy relationship between the benefits of ecosystems for humans and the integral benefits they have outside of anthropocentric concerns has been part of a larger conversation taking place within environmental discourse. The battle over how best to preserve natural environments has been split between two camps, one advocating conserving natural resources for sustainable usage by humans, and the other preserving them for their own sake.

This debate was instrumental in the establishment of the early environmental movement in the United States, when naturalist John Muir and forester Gifford Pinchot clashed over their ideas of protection for natural environments on federal land; their visions of how this should be managed differed. Pinchot took a conservationist view, believing that if these lands were left solely for recreational purposes they would deteriorate, and that the only way to prevent this would be ensure they were economically valuable through activities like mining, logging, scientific research and hunting, which could be regulated and managed. Muir, however, believed that pristine environments should be protected from any exploitative human activities in order to let them flourish in their own right. While both parties achieved their goals to a certain extent, they both reflected the colonial mindset of the period, neglecting to take into account the former occupation of many of these lands by indigenous people, whose cultural traditions offered alternative approaches to living with the land.

The binary split continues today, and two recent initiatives reflect the ongoing difference in the conservation and preservation approaches. Spurred on by conservation organizations in 2019, a United Nations working group publicized a proposal to protect 30 per cent of the planet by 2030 through various forms of environmental management, increasing to 50 per cent by 2050. These global targets, using 2018 data as a baseline, intend to protect ecosystems and habitats featuring high biodiversity, and are set to be ratified by the Conference of the Parties to the Convention on Biological Diversity.

The other project, spearheaded by E. O. Wilson, calls for half the Earth to be preserved, effectively by ring-fencing biodiversity hotspots around the globe, as human-free zones to protect them in perpetuity. It is based upon Wilson's Island Biogeography ecological theory, which predicts that 'a change in area of a habitat results in a change in the sustainable number of species by approximately the fourth root,'[1] meaning in terms of species diversity that the larger the area of land set aside, the greater the benefit. He calculates that 85 per cent of species will be protected by this means, ensuring a safe level of global biodiversity for the future. As a solution to the existential problem of mass extinction, it comes into conflict with political and social justice issues, and one difficulty with the project is working at national and local levels of governance to ensure that effective buy-in can be achieved in an equitable manner for both people and the environment, respecting the needs and integrity of each. Biodiverse areas will be created from a network of core sites, such as national parks, reserves and restored landscapes, chosen for the amount of richness of species they contain and linked together by corridors. These areas will be ecologically or biologically significant, intact ecosystems and ecosystem services, including those that address climate mitigation through carbon sequestration.

While conservation and restoration debate the different merits of human relationships to existing functioning ecosystems, the practice of restoration runs parallel to them, addressing environments already degraded by human activities. The aim of restoration ecology is to return these degraded ecosystems back to a previous healthy state. This remedial work can entail restoring the original composition of an ecosystem in an act of fidelity that reflects the history of the place, or it can be a functional restoration, prioritizing the quality of its health and integrity of operation over species composition. Working with vegetation, it means determining a selection of species for a plant community, then facilitating and ensuring its growth to achieve a predetermined target state. The problem with this approach has been that it assumed an ideal previous state to which it should be returned. Exactly what this target state is can be a matter of contention. Generally, it is considered to be a point before human degradation or occupation, but that point can be based upon assumptions that may be unknowable or affected by bias, given that human influence has affected most landscapes throughout history to some degree or another.

A pioneering early project of modern restoration ecology in the United States took place in 1936 at the University of Wisconsin–Madison Arboretum under the guiding eyes of ecologist Aldo Leopold and plant ecologist John T. Curtis. Civilian Conservation Corps workers reintroduced tallgrass prairie species to former arable land to recreate a Midwestern prairie (Plate 9). However, this assumed natural target state was far from natural, as the team had neglected to account for long-established indigenous management practices such as burning, which had maintained the ecosystems in the states they were prior to colonization.

Given that evolution progresses with time, the rationale for a predetermined target state does not necessarily mean that it would still be the most appropriate for now, or for changing conditions in the future. As successional processes are unpredictable and can evolve into multiple potential states, modern restoration practices often set an intended range of end states rather than just one, with the main objective being to achieve a healthy state.

Rewilding has become the latest fashionable iteration of restoration, covering a wide-ranging spectrum of interpretations and approaches driven by both ecological science and cultural values. The initial concept was developed within the conservation arena in 1998, when conservation biologists Michael Soulé and Reed Noss proposed it as a method that 'emphasizes the restoration and protection of big wilderness and wide-ranging, large animals – particularly carnivores'.[2] A recent attempt at a standard international definition has been proffered by the IUCN (International Union for Conservation of Nature) Commission on Ecosystem Management Rewilding Thematic Group, in which they describe rewilding as

> the process of rebuilding, following major human disturbance, a natural ecosystem by restoring natural processes and the complete or near complete food-web at all trophic levels as a self-sustaining and resilient ecosystem using biota that would have been present had the disturbance not occurred . . . Rewilded ecosystems should – where possible – be self-sustaining, requiring no or minimum-intervention management (i.e., *natura naturans* or 'nature doing what nature does'), recognizing that ecosystems are dynamic and not static.[3]

While a lot of rhetoric around rewilding is veiled in romanticism and cultural nostalgia, actual projects vary in their scope of expectations and scale of actual endeavours. The distinguishing factor between them is often similar to the debate within traditional restoration: the definition of the target state. The idea is that the baseline is the point before anthropogenic disruption of the ecosystem, which can be fairly recently or, in some cases, an idealized point in the past. The rewilding approach often involves the introduction of keystone species that may have been present in the original trophic chain at the site, such as apex predators like wolves. Comparable surrogates are also employed, where legislation permits, to replicate missing ecological roles, such as large bovine species. Popular substitutions for bison, aurochs, wild horses and boars are often deployed in European projects. In this respect, rewilding has lower fidelity to taxonomic precedent than restoration ecology, and can at times be used as a USP for safari-park-type experiences, hunting reserves or fashionable rare-breed farms. Some of these clearly negate the guiding principle of employing only the most minimal management interventions and use of non-predators implies stewardship, reinforcing yet again anthropocentric dominance over the rest of nature rather than promoting a new appreciation of it. Others are more forward thinking, embracing the opportunity to use the process as a means to research how ecosystems reassemble when left alone, using comparative studies to match expected development with actual progress, in order that we can learn from them rather than simply control them.

NOVEL ECOSYSTEMS

Beyond the preservation, conservation and restoration perspectives, scientific developments in ecosystem studies have led to new ways of thinking about the environment, and particularly the role of plants. The realization that humans have affected nearly every part of the planet, whether intentionally or not, has led to a new view beyond those inspired by pristine ideals or those programmed for profit. Rather than fighting an incessant battle to preserve, conserve or restore, the need to find an accommodation with the changes that are taking place, and to work with them pragmatically, has produced an appreciation for what are known as novel ecosystems, which have no previous analogue or reference points.

New assemblages of species that occur as a result of human activities, climate changes and intentional or inadvertent introductions from other regions have created ecosystems that are as biodiverse, functional, resilient and self-sustaining as many traditional ones. From an ecological point of view, ecosystems change in response to disturbance and other external factors, and species compositions vary through time. At some stage all ecosystems were novel, so in a sense the term could actually be considered redundant, but given the recent acceleration of change in the environment due to anthropocentric activities, the rate of unique assemblies of species is becoming more frequent, as species migrate and continue to be redistributed by humans, intentionally or otherwise.

Novel ecosystems can have a lot to offer, as they are adapted to unique situations and often more accommodating of change, but for many people they are disconcerting because they lack comforting cultural reference points and do not conform to expectations of national or regional vegetation typologies. The debate concerning the benefits of native versus introduced species has been a particularly vociferous one, especially in relation to plants. The idea of native species developed from the work of British botanist John Henslow in 1835 to distinguish specific British flora from neophytes. Over a century later, Charles Elton's 1958 book *The Ecology of Invasions by Animals and Plants* set the scene for what was predicted as an upcoming environmental apocalypse, with native species being overwhelmed by those from elsewhere. The deployment of military metaphors ever since has not helped to engender sensible discussions regarding the issue.

The pro-native argument proposes that plants have co-evolved to be suited to the specific environments they occupy, in relation both to biotic and abiotic factors. There is an evidence base to substantiate these points, but the percentage of symbiotic specialist species compared to generalist ones is actually quite low. Also, native vegetation may be a sign of evolutionary adaptation to local conditions, but it is not a guarantee that it is the best adaptation for an environment or that it provides the most beneficial effects. As evolutionary biologist Stephen Jay Gould noted,

'Once a species prevails over others at a location, no pressure of natural selection need arise to promote further adaptation . . . the great majority of successful species are highly stable in form and behaviour over long periods of geological time – not because they are optimal, but because they are locally prevalent.'[1]

A strictly native bias also denies the ecological reality of plants and other species, which have always moved around and naturalized. Most of the time this is not necessarily a problem, but concerns are raised when a species becomes rampant and is deemed 'invasive'. Ecologically these species are simply doing what they have always done, adapting to new challenges in order to survive, and in fact doing it very well. The moralistic overtones of the language of invasion anthropomorphizes these species, implying that they are acting consciously with intention. Invasive species do indeed cause disruption, and much of the negative press they receive is due to social and economic consequences, which usually have good reason to be addressed, but the conflation of these with ecological damage muddies the water.

When introduced species occupy certain habitats they can certainly disturb the existing community balance, creating a series of knock-on effects. Understanding these effects is key; it is not a matter of simply suggesting that this is a consequence of the ecosystem losing integrity. Functioning ecosystems are important for their dynamic and complex processes, which are related to their diversity, and should be the basis for considering any responses.

One thing is certain: there is no possibility of turning back the clock to return every ecosystem damaged by human activity to previous states. In the age of the Anthropocene, we are faced with coming to terms with the new associations of plants and other species that surround us. With this in mind, conciliation biology takes the approach that trying to eliminate or prevent invasive species is an unrealistic task, and instead sets out to monitor and manage outcomes of interactions between native and introduced species. As a strategy of coexistence, evolutionary biologist Scott Carroll suggests that 'conciliatory strategies incorporate benefits of non-natives to address many practical needs including slowing rates of resistance evolution, promoting evolution of indigenous biological control, cultivating replacement services and novel functions, and managing native–non-native co-evolution.'[2] So long as we work with rather than against them, novel ecosystems can be potential allies in creating resilient and adaptive environments in the face of a changing climate. These communities 'may also be more susceptible to proactive eco-evolutionary manipulation than the more integrated and redundant structures of deeply coevolved native communities'.[3]

Taking things a bit further, ecologist Michael Rosenzweig's concern that there is not enough available land to successfully preserve biodiversity led him to develop the concept of reconciliation ecology. In its practical application, the idea actively attempts to increase the possibilities for more diversity to thrive: 'Reconciliation ecology discovers how to modify and diversify anthropogenic habitats so that they harbor a wide variety of wild species. In essence, it seeks techniques to give many species back their geographical ranges without taking away ours.'[4] This win-win approach has found particular resonance in the agricultural community, where strips of land can readily be given over to habitat creation without compromising productivity of the land.

URBAN ECOLOGY

Most ecological approaches to the environment are premised on ideas of either pristine or carefully managed landscapes, but the most novel of ecosystems are to be found in the spontaneous coalitions of native and introduced species cohabiting in cities. Like people, many of these have arrived through migrations from near and far, and the absence of any enduring native habitats provides a blank slate for new lives with varied neighbours. It is precisely the influence of humans that creates these environments, thereby providing the preconditions for novelty. The discipline of urban ecology developed to address the relationships between all organisms, human and more-than-human, within this distinctly anthropocentric environment, and given the fact that more than half of the human species in the twenty-first century live in cities, understanding them is extremely important for developing adaptive strategies for the future. Urban ecology takes cities as a natural form of human habitat, functioning like any other environment, looking at the relationships between human and other species' activities in a holistic way.

The discipline, which developed around early ideas from the Chicago School of Sociology in the 1920s in which ecological concepts were used to study the patterns and dynamics of cities, has evolved, drawing on the methodologies of other fields of study including geography, anthropology, psychology, economics, planning and architecture. Drawing parallels between materials, energy flows and processes, urban ecology investigates the relationships between them in both organic and cultural terms. The complex interactions between ecological, economic and social functions can be identified with their own spatial organization and patterns across many scales, and changing through time, affecting all species' behaviours, populations and community dynamics. Studies either consider the way that these are going on in specific ecosystems within a city or view the city as a whole ecosystem in itself. One perspective in urban ecology has orientated itself towards using social science methodologies in order to make cities more sustainable. Influenced by the Bruntland report, *Our Common Future*, and the UNESCO Man and Biosphere (MAB) programme, launched in 1971 to improve relationships between people and their environments, it is normatively orientated towards design, planning and policy. Another, more traditional, natural science strand is less value-based, concerned with the biological life in cities and how it is impacted by politics, urban development and other social activities.

Plants are central to the latter approach, as the dense fabric of the built environment and its peri-urban surroundings provide a wide range of potential niches for spontaneous ruderal species to occupy, revealing their tenacious tendencies to survive anywhere within their ecological tolerances. These random assemblies of vegetation, finding refuge in sidings, industrial sites, pavement cracks and brownfield sites, and collectively constituting what Richard Mabey dubbed the 'unofficial countryside', often exhibit a greater degree of diversity than many parts of the official version, and offer havens and sustenance for other city-dwelling life forms.[1]

As a distinct type of environment, the urban realm is ecologically disparate and geographically fragmented. What were once originally functioning features may have been destroyed, diverted or subsumed. Over time, rivers have been culverted or piped underground, marshes drained, woodlands segregated into isolated stands, and fields built upon. The continual loss of these to the built environment to accommodate ever-increasing urban density has consequences for the creatures that support and are supported by them, including humans. But these are the conditions that make urban landscapes what they are and the ecologies within them unique. Urban botany began to develop in the 1930s and picked up pace a few decades later in some of the devastated postwar landscapes of European cities.

Researching the vegetation in the wastelands of Berlin, Herbert Sukopp's 1979 seminal article, 'The soil, flora, and vegetation of Berlin's wastelands', showcased the uniqueness of the ruderal flora, much of it introduced, growing alongside native plants in the rubble sites around the city. Sukopp's work challenged the received wisdom that urban plant distribution patterns were merely coincidental and not worthy of serious consideration. Instead, he revealed the city as a new type of environment, which needed to be appreciated in its own right: 'Urban biocoenoses are an extreme example of communities produced by successive invasions and not by co-evolutionary development. In principle, the historic uniqueness of urban ecosystems, i.e. their combination of environmental factors and organisms, differentiates them from most natural ones, even those subject to strong disturbance.'[2] This uniqueness in his opinion necessitated a new view in which 'urban areas cannot be seen as ecologically homogenous with regard to climate, soil, and buildings. Rather, a mosaic of different biotope types, dependent on the small-scale distribution of land uses, can be distinguished. These biotopes can usually be clearly separated, and are themselves relatively homogenous. "Natural" ecological conditions are outweighed and modified by the type of land use in an area.'[3]

Increasingly, ruderal vegetation in cities is gaining recognition for its ecological importance, as are the more obvious areas of green space provision, such as parks and reserves, for their myriad benefits (particularly since the Covid-19 pandemic). Other less obvious parts of the urban environment are also benefiting from the power of plants. In the past decade there has been a crossover between planning-based urban ecology and its plant-orientated sibling. The term 'green infrastructure' describes strategies for using plants to address many of the urban problems that exist as legacies of poor planning, as well as creating mitigating and adaptive measures for future climate scenarios. Its practical application aims to create or restore ecological functioning within the built environment by using plants' natural processes.

Planted areas are important in creating biodiversity corridors linking up fragmented urban green spaces, providing important transit routes and habitats for many creatures, as well as for invertebrates and pollinators. With space often at a premium, additional biomass can often be added with green roofs and living walls, some of which are more effective and sustainable than others, depending on the methods used. Vertical greening is becoming more popular and is often incorporated in forward-thinking architecture. Singapore has been a front runner in showing the diverse ways in which plants can exist above ground level and leading the charge on turning the urban grey to green.

Strategic tree planting is used to create urban forests, instrumental in locking up carbon, and assisting with air filtration by removing atmospheric particulates. Trees also provide cooling shade which reduces the 'urban heat island effect', whereby, due to the preponderance of hard surfaces and heat-generating everyday activities, heat is trapped between buildings, raising temperatures above those of surrounding areas. Conversely, as wind buffs, trees also help to mitigate cold weather.

Planting performs water management services, packaged under the acronym SUDS (Sustainable Urban Drainage Systems), by regulating the amount of rainwater run-off from roads and pavements to drains through absorption into the soil, helping to prevent flooding. Design solutions can assist in this by directing run-off to planted areas or through specific bioswales, large vegetated channels to which excess run-off is channelled to dissipate, while also removing debris and pollution.

While utilitarian in application, the creative potential for working with plants in cities transforms them from spaces to places, making them more amenable for people to inhabit and enjoy, as well as helping with health and well-being. All of these interventions work alongside the existing pockets of personal and communal greenery that surround homes, which also have their own roles to play in contributing towards a more biodiverse future.

GARDEN ECOLOGY

While gardens are certainly cultural creations, they are nonetheless ecologically important environments and popular expressions of novel ecosystems. After all, gardening involves the curation and construction of specific local ecologies of varying sizes that link up to neighbouring gardens, parks and streetscapes to create networks of biodiversity corridors through the built environment. Regardless of their state of cultivation or neglect, they all have something to offer a wide range of species that play different roles in these dispersed ecosystems.

In England, residential gardens occupy 633,000 hectares/1.5 million acres.[1] While the spread is not socioeconomically or ethnically balanced, the majority of dwellings have access to an outdoor plot of some kind.[2] Their importance to biodiversity has traditionally been downplayed and wasn't really given a thorough grounding until the late twentieth century.

Some of the most important work done on garden ecology was carried out in Leicester by the academic zoologist Jennifer Owen. She devoted three decades, beginning in 1972 and ending in 2001, to assiduously observing her domestic plot in order to reveal its richness as a thriving biodiverse environment. She managed to identify 2,673 different species living in her 741 sq m/886 sq yd plot, including 474 plants, 1,997 insects, 138 other invertebrates (such as spiders, woodlice and slugs), and 64 vertebrates (including 54 species of birds and seven mammals). Her early findings were published in 1983 in *Garden Life*, followed by numerous journal and magazine articles, culminating in *The Ecology of a Garden: The First Fifteen Years* in 1991, while a final thirty-year summary, *Wildlife of a Garden: A Thirty-Year Study*, appeared in 2010.

Owen recognized the artificiality of gardens, and the conscious collection of plants within them, not only as something quite different from natural habitats, but as something uniquely valuable: 'I have called this diversity contrived diversity because it is achieved by the decision of the gardener to grow as many different plants as he can fit on to his plot. Aliens originating in many different parts of the world, together with native plants, are cultivated side by side . . . This extraordinary assemblage of species, diverse taxonomically as well as in country of origin, is typical of the plant diversity that a gardener contrives in a small space.'[3]

Owen also recognized the importance both of the intentional plants tended by the gardener, be they cabbage or cornflower, and of those unintended plants that somehow made their own way there and found the conditions amenable, such as chickweed and oxalis. Her own garden harboured 412 plant species between 1975 and 2001: 311 were cultivated and 101 came in of their own accord. In 1984 alone, it contained 264 flowering plants, excluding grasses, of which 197 were deliberately cultivated, producing an astonishing diversity of 3,563 species per hectare/2.5

acres. Her embrace of them all celebrates the contributions they make to the bigger picture: 'Since green plants are the starting points for all terrestrial food chains, and many, perhaps a majority, of insect herbivores are confined to one food plant species, or at least a group of related species, diversifying the plants generates many different food chains.'[4]

Inspired by Owen's research, the Biodiversity in Urban Gardens Study (BUGS), steered by the University of Sheffield, undertook a comprehensive study of gardens, looking at plants and other forms of biodiversity, which confirmed many of her findings. The first part of the project involved surveying gardens in Sheffield in 2002, followed by a national reconnoitre from 2004 to 2007. The project studied the composition and distribution of garden floras, placing them in the context of semi-natural and other urban floras. It found that gardens contained remarkable levels of floristic diversity, with a higher number of species than in other types of landscapes, and featured a ratio of 30 per cent native species to 70 per cent introduced. This richness was attributed to two factors:

> First, there is a very large pool of plants available to gardeners – the current Royal Horticultural Society Plant Finder (Macaulay et al. 2002) lists over 70 000 taxa (*c.* 14 000 distinct species) available from UK nurseries. Second, owing to active management and maintenance by gardeners, garden plants have a highly 'unnatural' ability to persist at remarkably low population sizes. Another unusual feature of gardens, relative to semi-natural habitats, is that gardens are much less homogeneous, i.e. individual gardens may be largely open, grassy, or deeply shaded, while larger gardens may contain all these habitats.[5]

The results were consistent across the cities studied, revealing that the positive benefits plants have to provide is something that is determined more by culture than by climate.

Despite the growing stature of the environmental movement throughout the 1970s, it had little impact on mainstream publications in the gardening world. The fact that horticulture and ecology were seen to have nothing in common was something that Stefan Buczacki attempted to address with the publication of *Ground Rules for Gardeners: A Practical Guide to Garden Ecology* in 1986. Aimed at the average amateur gardener, it laid out the basic principles of ecology in a straightforward manner. Buczacki attempted to encourage gardeners to consider the bigger picture of the context and processes which gardens were a part of. His follow-up book, *Garden Natural History*, published in the influential New Naturalist Library series by Collins in 2007, pursued the themes of his earlier work with more forensic intent, updated with more recent data and driven by what he still saw as a disappointing lack of correspondence between the disciplines of ecology and horticulture.

Calls from academia have also encouraged urban dwellers to embrace their gardens, and to appreciate the anthropogenic jumble of urban plant assemblages as being of intrinsic worth. Professors James Hitchmough and Nigel Dunnett have questioned the devaluation of intentionally created landscapes:

> Why, for example, should plant communities brought into effect by intentional (or unintentional) human agency be ecologically and aesthetically intrinsically less valuable than those that result from random combinations of chance events? In biological terms they may be demonstrably less or more valuable, depending on their architecture and the species present, while in most cases being more aesthetically pleasing due to having been so designed. Why is human agency so bad when it was unconsciously or consciously employed in the past to help create semi-natural vegetation, such as meadows, steppe, prairie, and various woodland communities that we now cherish as 'nature'?[6]

As the research reveals, the ecological value of the management of gardens is an important tool that can be used to increase biodiversity and ecosystem functions in the face of the climate crisis. The question is: how can these benefits be maximized most effectively, and are gardeners up to the challenge?

CLEANING UP THE GARDEN

Gardeners will always be the first to say that they, of all people, don't suffer from any such condition as a lack of plant awareness or aversion to the vegetal world, and many self-professed plantaholics will be quick to stress that their activities are beneficial to themselves and the world around them. Indeed, they are, to some extent, but is the idea prevalent in the garden world that 'green is good' really the case, or are the actions of gardeners actually still part of the problem rather than the solution? The industrial scale and cycles of production and consumption in the horticulture industry are so far removed from any forms of ecological processes as to suggest the opposite. Perhaps this is unsurprising, given that horticulture's roots lie in its historical development as an offshoot of agriculture, the paradigm-shifting technology that marked the nascent stirrings of the disconnection of our species from the rest of the natural world.

Agricultural production, based upon the domestication and modification of plants, is carefully attuned to maximize yield. This is achieved by divesting ecosystems of their complex networks and the life forms that inhabit them, something clearly visible in the huge swathes of monocultural crops that cover the arable parts of the globe today. Isolating crop species in this manner improves efficiencies by transforming minimal resources into maximal future assets. While the cost-benefit analysis of this is valued in economic terms, the loss of biodiversity is simply written off as worthless collateral. Large-scale horticulture operates on the same premise, and the association between the two industries is clearly evident in the close alliance between trade bodies. There is clearly no sentimentality in horticulture on this level, no attachment to the beauty, benefits or wonder of plants; they are merely livestock products.

Perhaps most important is the fact that the industry is still firmly wedded to the unholy trinity of pesticides, peat and plastic pots, despite the fact that it is now widely known that all have negative environmental impacts. As in agriculture, pesticides are routinely used to kill any threatening biodiversity when growing commercial plants. The development of pesticides for use when growing crops has a disturbing history in the production of chemicals for warfare.[1] In 1962, Rachel Carson pointed out the devastating ecological effects of DDT use in agriculture in her book *Silent Spring*.[2] More recently, the negative health effects of glyphosate and neonicotinoids, for humans and other species, have been highly contested and their knock-on effects for biodiversity well documented, yet they are still widely used.[3] A few progressive commercial growers are now using biological controls, but this is not yet mainstream.

Peat is used as a growing medium for its nutrient content and water-retentive qualities, which help accelerate growth and maximize yield – and therefore profits.

Peat wetlands are unique ecologies that store carbon from decayed plant matter, which is released as CO_2 into the atmosphere when it is dug up, accelerating global warming. Very specific flora grow in these unique habitats, so the idea of cultivating ornamental plants that have no relation to these environments in peat-based media is a highly unnatural and unnecessary process. As awareness has grown over the detrimental effects of peat extraction and its contribution to climate change, there has been a move towards alternative media, such as organic matter, green waste and coir, but the uptake has been slow. The voluntary reduction policy established by the UK's Department for Environment, Food and Rural Affairs (DEFRA) in 2011, intended to eliminate usage by 2020, failed due to a lack of incentive to achieve a zero target. Two and a quarter million cubic metres of peat were sold for horticultural use in 2020, approximately 30 per cent to commercial growers and 70 per cent to domestic consumers.[4] At the end of 2021, DEFRA announced a consultation to seek views on measures designed to end the sale of peat and products containing peat in the retail sector in England and Wales, and to call for evidence on the impacts of ending their use in the professional horticulture and wider sectors.[5] The proposed measures under consultation only apply to bagged growing media and not ornamental nursery stock. Despite these limitations, the announcement sparked a heated debate within the industry, with large-scale nurseries resisting any form of change, fearing that their profit margins will be impacted. Smaller nurseries, however, have shown a willingness to adapt, despite the fact that due to economies of scale, it will be more expensive for them to reorientate their production than the larger growers. Bridging the gap between the industry and domestic gardeners, the Royal Horticultural Society have committed to be completely peat-free across all their operations, including gardens, shows and retail, by 2025.[6]

Single-use plastic pots are the primary containers in which plants are grown and sold in the UK. Standard black pots are not recyclable, as the machines at recycling depots are unable to identify them due to their non-reflective colour – and so 300 to 400 million of them go to landfill or incineration in the UK each year.[7] In the USA around 4 billion pots that are produced annually are condemned to a similar fate.[8] Recently, taupe-colour pots with recycling potential have been introduced, but they have yet to achieve widespread use within the industry or acceptance by local authorities. Other biodegradable pots have been on the market for some time, but are only used by smaller independent growers, businesses often run by individuals or families who, despite the disproportionate higher costs for them, are prepared to take extra measures for the sake of the environment.

Assuming that these problems are addressed in the near future – and they could easily be with effective policy implementation and oversight, and positive encouragement of behaviour change from the media – the next step towards environmentally friendly gardening is to rethink our attitude to plants beyond purely aesthetic concerns to more holistic ecological approaches that reorientate how we inhabit and enjoy our green spaces.

PLANTS AS PICTURES

GROWING THE IDEA OF THE GARDEN

Gardens are an important form of connection with nature, providing ongoing engagements with other organic life forms and the natural processes they are part of. However, they are also a form of disconnect, shaped by all of the social factors that construct our ideas of nature and sense of separateness from it. They are an arena in which our instincts of biophilia and our desire for control are played out in an ongoing battle, and the resulting spaces produced by this war of attrition are unconscious expressions of these complex and contradictory urges.

As unique biocultural creations, gardens are distinctly human responses to geography, geology, topography and climate, shaped by social needs and aspirations, and realized through the co-option and cooperation of other life forms, most notably plants. They provide physical and mental nourishment and connections with the wider landscape and give a sense of rootedness to place. As products of environmental as well as social conditions, they have materialized more often where climates are most favourable, and where there has been a cultural desire for a mediated connection with the rest of nature. Their popularity can be seen throughout history, from their earliest iterations in Sumeria, Mesopotamia and Persia, through Chinese and Japanese gardens that created dialogues with the wider landscape, Islamic gardens conceived as expressions of paradise, to the European tradition, which has been exported through colonialism and migration to the Americas, Australia, New Zealand and beyond.

It is not simply private residential gardens but also other spaces, such as parks, community gardens, allotments and streetscapes, where we cohabit with plants. The private realm speaks volumes about personal values, aspirations and senses of identity. Gardens are expressions of all of these in differing ways, some very obvious, others less so. The fact that access to gardens is skewed by class and ethnicity reveals issues of both social and environmental inequity. Public spaces also embody aspirations but on a civic, communal level. As part of the body politic, the diversity of demands that these democratic domains are required to meet is continually contested, reflecting shared values and the ways that we see ourselves as groups, large and small. Their designs, and the plants within them, say as much about us as do the statues they home.

The connections that gardens provide with the more-than-human world are managed in a contained and controlled way. The philosopher Francis Bacon, in his 1597 essay 'Of gardens', defined the garden as an escape from the threats of nature, and the traditional enclosed garden form, the *hortus conclusus*, was intended to keep the wilderness at bay yet contain a refined version of it within its bounds, to be carefully protected and nurtured – an idea that is still the basic blueprint for gardens today. As safe and secure spaces that mitigate the vicissitudes and threats of the external world, they are part of our habitat, but unlike architecture, which aims to be solid, static and enduring, gardens are the opposite: ever changing, often unpredictable and, despite gardeners' best efforts, not completely under their control.

Gardens are places in which part of our long and entangled co-evolution with plants has been, and continues to be, played out. They take pride of place, providing visual pleasure, stimulating the senses, enhancing health and well-being, and embodying symbolic meaning. The domestication of plants began before the pragmatic practices of agriculture, initiated by a sense of enchantment and wonder at plants themselves. Plants in gardens are not solely orientated towards sustenance and survival, but are there by dint of the fascination they exert over us. Although productive growing has been an important part of garden history, whether it be the vegetable plot, kitchen garden or allotment, it is the cultivation of plants for their aesthetic appeal that has over time come to define gardens as unique and distinct from other types of spaces.

Garden historian John Dixon Hunt describes gardens as a type of 'Third Nature': 'Those human interventions that go beyond what is required by the necessities or practice of agriculture . . . some relative elaboration of formal ingredients above functional needs, some conjunction of metaphysical experience with physical forms, specifically some aesthetic endeavour – the wish or need to make a site beautiful.'[1]

The Hebrew term *gan eden* evokes the garden as a place of well-being and joy. In the biblical story, the Garden of Eden was free from the burden of toil, somewhere for relaxation and enjoyment. It was a place of abundance, satisfying both nutritional and emotional needs. This idea of the garden as paradise has filtered down from Islamic gardens to permeate still the ways in which gardens are thought about today, as places of leisure and pleasure. Of course, the irony of gardens as places of pleasure is that to create and maintain them in their ideal states is far from easy, and actually involves a great deal of labour. The dialectic of enjoyment and exertion is relived over and over again by gardeners with every shove of the spade and every snip of the secateurs.

But the idea of gardens of leisure, freed from the concerns of material provision, represents a form of privilege, a step on to the next level of the hierarchy of needs, as garden historian Derek Clifford suggests:

> It is no good looking for gardens in a society which needs all its energies to survive. As soon as a society has time and energy to spare, some of the excess is devoted to enjoying the residual aspects of enclosure, of cultivation, and of unhumanized landscape. The way in which that residue is shaped to give pleasure depends partly upon the physical opportunities, but far more upon man's spiritual needs. A garden is man's idealized view of the world; and . . . it follows that fashionable gardens of any community and any period betray the dream world which is the period's ideal.[2]

Leisure implies a certain level of disposable time and income. For much of garden history, it is the gardens of the wealthy and powerful that predominate, in the same way that their lives dominate social history. These gardens are products of privilege, and it is not surprising that many were created in religious, royal, aristocratic and imperial contexts until the growth of middle-class consumerism in the Global North, at the end of the nineteenth century, began to spread garden ownership across a broader demographic.

Tied up with privilege are expressions of taste: sensibilities built around knowledge and status exhibited through lifestyles, behaviour and possessions. A social asset closely related to – but independent of – wealth and class, taste is a form of cultural capital that can be seen to run through the feted gardens of the past to those of the present. As a social statement, it reveals the arbitrary but influential benchmarks of what is and isn't considered appropriate in gardens and the ways in which plants are selected and arranged in them.

While planting styles display this influence in different ways, there are certain things common to all gardeners' aspirations. Planting is always a managed situation, aiming at achieving a state of arrested development that prevents the natural processes of succession from taking place, encouraging a sort of subclimax rather than letting plants take their own courses. For various obvious reasons, most gardeners are not keen on small-scale forests, woodlands or scrubland forming in their back yard. Rather they have in their minds a clear picture of a relatively static, picture-perfect ideal consisting of varying mixes of forbs, grasses, shrubs and trees in an apparently harmonious balance. This requires planting that is an ecological simplification of natural vegetation, and maintenance that aims to reduce competition and eliminate any intruders that fail to fit the preferred aesthetic criteria.

The balance between the ecological and the aesthetic in planting design presents a spectrum of different approaches, from those that aim to enhance artifice to those that emulate and even encourage the ecological. At one end, gardening has been about showcasing mastery by amplifying the human-orientated pleasure to be derived through the dominance of plants. Unabashedly anthropocentric, this prides itself on neatness and forcing plants to perform in ways that please people regardless of any natural tendencies they may have themselves, and it requires intensive input in terms of labour, fertilizers and pesticides to succeed. At the other end of the spectrum is a looser, more relaxed approach to gardening which seeks to delight in the very being of plants, and the benefits they may bring to other forms of life, requiring lower resources and intervention.

Playing out along this spectrum are tensions about how gardens should look, degrees of control and freedom to function without intervention, as well as which plants are welcome and which are not. A big part of gardening involves determining the latter, for it is not just any plants that are allowed into these gracious spaces; it is a matter of careful calculation and curation that bestows the privilege of belonging on certain species. Plant prejudice is often quite arbitrary but, more often than not, it is a product of consideration based largely upon expectations of the ways in which plants are required to look and 'perform' in gardens. Some features, such as big, bright, bold flowers and graceful forms, are coveted attributes for many gardeners, but above all else plants must be well behaved, suitably domesticated.

The garden wall is the line demarcating the domesticated plants within the garden and the wildness of natural vegetation outside it, a physical enactment of the nature/culture divide. It is a reminder that wildness is threatening, destabilizing and to be kept at bay, but it also acts to exclude the idea that plants can exist in themselves and for themselves, and are capable of living their lives quite successfully regardless of humans.

The fear of unexpected plants in a carefully tended plot reveals some of the cultural anxieties at the heart of gardens. Uninvited vagabond plants are considered devoid of aesthetic merit and disruptive to hard-won horticultural equilibrium, as nature writer Richard Mabey suggests: 'Plants become weeds when they obstruct our plans, or our tidy maps of the world. If you have no such plans or maps, they can appear as innocents, without stigma or blame.'[3] And it is because gardens are very much mapped and planned spaces, psychic strongholds and fortresses against the unfamiliar, that so much effort is expended to maintain a tight roster of residents. If one were to believe the vast body of horticultural literature that has accumulated over the last century, then one would be assured that 'weeding' is both a moral and a practical imperative upon which the fragility of the perfect garden depends – an insistent activity that is rather ironic given the origin of gardens as places free from the necessity of toil.

Domesticated plants without strict utilitarian value, and those used in a purely decorative way, represent a luxury, something that nourishes aesthetically, emotionally or spiritually rather than fulfilling base needs. Gardens are privileged precincts that harbour these plants, where they are usually tended with care as if they were pets or admired as if they were ornaments, shorn of any intrinsic value they may have in their own right. Gardens are stage sets for plants plundered from diverse habitats around the globe and reconfigured into audacious ecologies.

Novelty has also played a big role in plant selection, driven by the addictions of gardeners seeking rarity and unusual attributes. But, ironically, 'Novelty for its own sake became, as it has remained, a horticultural desideratum and as a result the history of gardening is, among other things, a continuing chronicle of new species appropriated from the wild.'[4] The blurring of the boundaries between the garden and the wild is something that has intermittently surfaced in the history of gardens, revealing the potential of new balances between the controlled and chaotic aspects of plants.

Throughout history these concerns about plant selection and arrangement have played a role in defining the characteristics of gardens. As products of particular times and places, these relationships with plants are touchstones to wider cultural ideas about the natural environment articulated through the language of horticulture. Plants appear in gardens for many reasons and in many ways, selected, arranged and managed based upon a variety of social and botanic criteria. They have been used for pragmatic purposes in order to create spatial division and definition, screen and frame views, direct the eye and movement, provide shelter and shade, and create surfaces. Often their features have been employed to express form and order through their management in geometric and architectural ways, hewn into hedges and tapered into topiary, arranged in avenues, rows and beds. They have been used to aesthetically enhance, either for the admiration and adoration of their intrinsic attributes or through their consciously arranged groups that aim to express personal creativity, reveal taste and expertise, and create mood and atmosphere.

While their uses are determined by such design intentions, their specific selection is according to their qualities and characteristics, such as size, shape, texture and colour, their life cycles taking into account their annual, perennial, evergreen and deciduous tendencies, their competitive and invasive behaviours or propensity to self-seed. Appropriateness to site is key to satisfying their resource requirements for light, moisture and nutrients. Availability and novelty are social factors that play a role, as are maintenance and the time and money needed to keep them in their desired condition.

All of these considerations have been articulated stylistically, with plants appearing individually, in monocultural masses, mixed together, intermingled, or in matrixes that highlight plant associations in varying degrees of artificiality or naturalness. Over the course of time, these different styles reveal distinct tendencies privileging function, form and flowers. In particular, over the past 150 years a favouring of the floriferous has been the defining trait of British gardens, consolidated in the idea of 'the English Garden'.

Flowers have always enticed and allured with their intricate charms, but their presence as the defining feature of gardens became ubiquitous, as if planting design is simply a matter of exterior flower arranging. And to an even greater extent, it is not even the flowers themselves so much as their colours which lead planting design. Floriferous plant 'pornography' still dominates the garden media, as if the flower is the only part of the plant of any importance, and this constantly reinforces the idea that plants are merely there to make pretty pictures. This not only objectifies plants but also reduces their importance and the magnificence of their multifarious abilities and benefits.

Fortunately, a discernible trend has been developing over the past thirty years, moving towards a more holistic appreciation of plants, looking at new ways of working with plants in more 'naturalistic' ways in private and public places, thinking beyond simply the visual appeal of flowers and their colours to consider the wider implications of their ecological interactions within their environments. Much of this has been aesthetically driven but it has increasingly become influenced by science, as advances in our understanding of the environment, ecology and the effects of the current crises we face have brought about new ways of looking at the world around us.

On the level of the general public, a never-ending stream of books, magazine articles and television programmes shout the wonders of 'wildlife gardening' aimed at attracting bees and butterflies (those charismatic invertebrates, the garden equivalent of cuddly megafauna, like panda bears and penguins, which hog the media limelight when it comes to focusing on biodiversity). Beyond this, a number of designers and plant specialists are working with plants as complex communities, to devise planting schemes that can be more resilient faced with unpredictable weather patterns, and require a lighter, less disruptive touch when establishing, and less intervention and resources, such as water, fertilizers and chemicals, to maintain. These communities can work to ensure soil health, detoxify polluted sites through phytoremediation, create habitats and assist biodiversity: a series of wins all round for plants and people. Many of these developments originated in the realm of conservation but have increasingly been adopted in the public realm on projects of various sizes and are now being applied in domestic gardens. The principles are scalable and the aesthetics customizable to ensure that they fulfil the different social needs in each situation.

Ecological planting is aimed at addressing the changing conditions plants and people face, now and in the future. The fact that traditional styles of planting still hold sway with most gardeners today is testament to their enduring appeal, but simply doesn't address many of the issues that are becoming more and more relevant. Casting a cursory glance over gardens of the past can provide insight into the differences between these new methods and traditional horticulture, as well as an understanding of how they each came to prominence and the factors involved in their formation and development.

Plants have long played a structural role in defining spaces within the garden and distinguishing it from the world beyond its borders. Hedging, parterres and topiary have been ways in which plants' natural growth tendencies have been manipulated to create forms ranging from the functional to the fanciful to suit human needs and whims.

Strong formal structure created from evergreen trees and shrubs helped to define the era of formal garden design that reigned in a procession of different styles, from the Italian Renaissance gardens of the fifteenth and sixteenth centuries to the grand designs of landscape architect André Le Nôtre in France in the seventeenth century. The calculated curtailing of plants' natural tendencies by deliberately imposed external order was considered the highest form of garden sophistication. Plants left to their own devices were thought to be shapeless and imperfect, a situation that could only be remedied by human action. In Le Nôtre's designs for Louis XIV, in particular at the Palace of Versailles, formality reached its zenith through ostentatious displays of power and conspicuous wealth articulated through the medium of meticulously defined structural planting. As a device to express control over the landscape, it proved effective in realizing its anthropocentric intentions, forcing geometric form over organic shapes, denying plants their spontaneous ways of expression by enforcing regimes of manicured maintenance, and limiting any form of ecological richness through the restricted range used. 'The severe simplicity of the design contrasts with the waywardness of nature. Shrubs are pruned into triangles, spheres or cones, or into the likeness of men or animals – into shapes, in any case, which could not possibly occur without human intervention. Nothing could be more autocratic, better calculated to display man's power over nature.'[5] Throughout this period, the gardens of the British aristocracy were influenced by the ideas and trends of their European cousins, with evergreen architecture providing suitable structural framing and ornamental embellishment of country estates.

Concurrent with these gardens, a vernacular tradition of gardening developed, born of the pragmatic concerns of the rural working class rather than aesthetic inclinations. Initially intended to fulfil basic needs, it began progressively to incorporate decorative elements. Taking their cue from early town and monastic gardens, in which plants grown for their herbal and medicinal uses eventually became valued and nurtured for their visual appeal, manual workers began to transform the areas outside their homes. Existing side by side in villages and facing the road, they were denied the full sense of enclosure usually associated with gardens, with their combined effect producing a sense of community in stark contrast to the walled properties of individual landowners. Mixing herbs with vegetables and fruit trees, these gardens satisfied the practical needs of their owners. Their informal mix of produce and ornamentals – increasingly complemented by wild flowers found locally – became a fine balance of the untamed and the cultivated. They quietly evolved into a semblance of order with a distinctly loose appearance that came to be recognized as the 'cottage garden', a form given little significance in terms of planting design until much later.

An important introduction to the English country house gardens was the herbaceous flower border (derived from the French word *bordure*, meaning edge), which afforded opportunities for more floriferous displays within the formal structures of the garden, with graduated layers of plants arranged according to height around parterres.[6] Borders in seventeenth-century gardens were narrow and thinly planted, but in the nineteenth century they developed the form and scale which came to define them. In 1829, William Cobbett in *The English Gardener* outlined the differences between traditional beds and the new borders:

> Flowers are cultivated in beds, where the whole bed consists of a mass of one sort of flower; or in borders, where an infinite variety of them are mingled together, but arranged so that they may blend with one another in colour as well as stature. Beds are very little in fashion now . . . but the fashion has for years been in favour of borders, wherein flowers of the greatest brilliancy are planted, so disposed as to form a regular series higher and higher as they approach the back part, or the middle of the border; and so selected as to insure a succession of blossom from the earliest months of spring until the coming of the frosts.[7]

This quiet insurrection in successional mixed planting with a floral focus would in the future shape planting design for many years, when it finally found popular appeal at the end of the nineteenth century. But it had to bide its time, as this initial appearance was quickly overshadowed by a home-grown revolution in design that renounced all that had preceded it.

The English landscape garden ushered in an era shaped by idealized notions of nature and landscape, and fuelled by the literary exhortations of Horace Walpole and Joseph Addison. Designers such as Charles Bridgeman began to discard the shackles of artifice and formal geometry in a liberating gesture, celebrating 'In all, let Nature never be forgot.'[8] Rather than the controlling tendencies of formality, design was instead enlisted to respect and make explicit the natural forms of the landscape, thereby improving it: 'Consult the genius of the place in all.'[9]

This work of environmental amplification required a suitably trained eye capable of understanding which elements of the landscape were suitable for accentuation. Realized at large country estates, these creations in which 'nature' was improved and made more pleasing to the eye were at the scale of parklands rather than that of traditional domestic gardens. Their importance was not horticultural but rather concerned the garden's relationship to the wider environment around it. Their outward orientation turned the traditional garden template of the inward-facing *hortus conclusus* on its head, preferring to embrace prospect rather than shelter in enclosed refuge. Their defining feature was the blurring of the line between the large sweeping swathes of grassland constituting the garden and the land beyond it; as Walpole described it in reference to the work of landscape designer William Kent, 'He leaped the fence and saw that all nature was a garden.'[10]

Borrowing the landscape was a key technique to extend the view beyond the property's limits, giving not only a sense of its extension into uncultivated land, but also a greater sense of ownership, making a statement about the owner's status. The invention of the ha-ha – a ditch to prevent animals from entering the property rather than using fencing – was a key device to keep unobstructed views across the domain and accentuate the effect.

Kent's landscapes were influenced by the landscape paintings by seventeenth-century French artists Claude Lorrain, Nicolas Poussin, Gaspard Dughet and Salvator Rosa, which invoked Arcadian scenes of pastoral bliss. These idealized scenes were far from natural landscapes, but rather romanticized versions of Greek agricultural lands which had been tended for millennia. Recontextualized in an English setting, the artifice was further highlighted by the fact that the long landscape views extending into 'nature' were actually of tracts of previously deforested farmland, heavily indebted to human intervention and shaped by wealth, political will and power over many centuries.

This agricultural influence resulted in these landscape gardens reducing plant diversity to large lawns interspersed with strategically situated trees, as individual specimens or in clumps. The trees were spared the clipping so popular in formal design and allowed to grow in less tortured ways to enhance their naturalness. If Kent can be credited with any form of horticultural innovation, it is by mixing conifers with deciduous trees, which was novel at the time.

Ironically, the naturalness of the gardens could only be achieved by a sleight of hand, as the lawns needed to be laboriously maintained to achieve their perfect look. Assiduous attention was required around the house, where it would often take as many as three men with scythes a day to cut half a hectare/an acre of grass, along with women who followed behind to collect the clippings. At some properties this was carried out as often as two or three times a week. Even after the invention of the lawn mower, scything often continued to be used for its precise results at the hands of skilled practitioners. Weeding was also needed to ensure a 'flawless' look, exacting an ecological toll by denying diversity.

Further away from the house such meticulous manicuring was somewhat less critical, as fine detail was not so noticeable, so control was achieved by grazing sheep to maintain these landscape scenes in a semblance of static timelessness. This form of gentle disturbance would undoubtedly have been beneficial to some smaller plants within the grass swards, but any such biodiversity was incidental and certainly not part of the intended plan.

This painterly approach to perfecting nature through its careful arrangement was, later in the century, to achieve the status of landscape design as an art form in its own right, through the skilled stylings of Lancelot 'Capability' Brown. His considerable portfolio of work helped to shape not only the parklands of large numbers of country estates but also national notions of the 'green and pleasant land', which have echoed on down through the years. His distinctive work distilled that of his predecessors into a clearly articulated form, a style recognized as the 'English landscape garden', and successfully exported to the rest of Europe, where it was adopted in a fashionable manner. As with Kent, the key feature of Brown's planting was primarily trees used as elements to construct the desired visual tableaux. These were harmoniously arranged, extending across the parkland to a shelter belt around the perimeter, all anchored in a sea of grass that rolled right up to the foundations of the house. His harmonious vistas favoured artfully constructed serpentine lines in tribute to the lack of naturally occurring straight lines in the landscape. Unfortunately, his sensitivity to the graces of the natural world did not generally manifest themselves in a light-touch environmental approach, as fields – and even villages – were known to be flooded in order to divert rivers and create lakes, in a manner akin to the insensitivity of national infrastructure projects like the HS2 railway route today.

The economics of these endeavours was noted by Thomas Jefferson prior to his presidential stint in the United States. He visited Britain in 1786, undertaking a tour of many of the most notable gardens of the time in the landscape style, with a view to collecting tips for the garden he wished to design at his residence, Monticello, in Virginia. Surveying not only the topographical and horticultural features of the gardens, he was also keen to count the costs required for the upkeep of such large and impressive grounds. After visiting Stowe, he recorded '15 men and 18 boys employed in keeping pleasure grounds' and at Blenheim '200 people employed to keep it in order, and to make alterations and additions. About 50 of these employed in pleasure grounds.'[11] Natural landscapes clearly came at a cost.

Brown's successor Humphry Repton, keen to quickly assume his mentor's mantle, maintained throughout his early work the basic blueprint of improving and enhancing landscapes with large-scale gestural brushstrokes by modifying their existing topography and framing views to pleasing effect. His later endeavours showed the influence of the debates around the Picturesque aesthetic movement, proposed by theorists William Gilpin, Uvedale Price and Richard Payne Knight, and influenced by philosopher Edmund Burke's treatise *A Philosophical Enquiry into the Origin of our Ideas of the Sublime and Beautiful* (1757). The style sought to eschew the polite forms of harmony and beauty advanced by Brown in favour of landscapes incorporating wilder features. Repton's response to the trend was minimal in design terms but did manifest itself in a more considered approach to achieving a sense of naturalness based upon close observation (Plate 1).

While Repton's designs occasionally incorporated flower gardens, they were to 'be an object detached and distinct from the general scenery of the place' and 'should not be visible from the roads or general walks'.[12] In these enclosures, 'rare plants of every description should be encouraged, and a provision made of soil and aspect for every different class.'[13] These features were clearly not considered part of the design proper, in which planting was all a matter of trees, such as oaks, beech and birch, and brushwood thickets of thorns, hazel, maple and holly.[14]

By employing a denser approach to planting trees than Brown, Repton created his own signature style, often using trees to conceal rather than reveal views, encouraging a sense of journey around the property. He planted them as if they had self-seeded, such that they should appear more naturalistic, but also as an acknowledgement that they actually grow faster when planted together, a perceptive insight at the time into the ways in which tree communities symbiotically support each other. Trees of the same species were incorporated into a group, often planting more than one in the same hole, suggesting different life-cycle stages, as if some form of succession was taking place: 'No groups will appear natural unless two or more trees are planted near each other, while the perfection of the group consists in the combination of trees of different age, size and character.'[15]

This appeal to naturalness was frowned upon by one of Repton's ardent followers, the influential garden writer John Claudius Loudon, who despaired that plants 'were indiscriminately mixed, and crowded together, in shrubberies or other plantations; and they were generally left to grow up and destroy one another, as they would have done in a natural forest; the weaker becoming stunted, or distorted, in such a manner as to give no idea of their natural forms and dimensions.'[16]

In reaction to this, in the early nineteenth century Loudon instigated the gardenesque style. This favoured an exaggerated gallery-style approach to planting, isolating specimen plants and elevating them to the status of art objects. Individual specimens or groups were arranged in the garden within scattered island beds or artificially raised mounds, such that their deliberate placement would allow them to be appreciated for their distinct qualities from a variety of different angles. The use of introduced plants was instrumental as it offered opportunities to highlight unusual forms, such as those of newly arrived plants like pampas grass from Argentina and monkey puzzle trees from Chile.

This obvious fetishization of plants in their singular glory, all suitably separated from each other, exhibited an almost zoological approach which, aside from any potential diversity of plants involved, offered little real ecological value. Instead it reflected a highly anthropocentric conceit towards using plants in gardens, neglecting their multiple facets in favour of an appreciation that can only be had when marvelling at them. Loudon was enthusiastically adamant that 'All trees, shrubs and plants in the gardenesque style are planted and managed in such a way as that each may arrive at perfection and display its beauties to as great advantage as if it were cultivated for that purpose alone.'[17]

Plants had long been collected and transported around the globe, but the sudden influx of plants that enabled Loudon and his acolytes to have such a fine array of specimens to display was a result of the expansion of British colonial activities in the seventeenth and eighteenth centuries. This period witnessed an incredible explosion in plants displaced from their native habitats and relationships with local biodiversity, cultural practices and knowledge.

The development of botany and Western Europe's imperial aspirations were closely allied through territorial expansion and commercial exploitation. The endeavours of the extractive industries, dedicated to removing plants from their natural habitats and homelands in the name of colonial botany, shared the same 'resource' mindset that also inhumanly uprooted people from their homelands during the slave trade. Initially, the plants sought out were those species that provided novel sources of revenue, such as sugar, nutmeg, tobacco, cloves, cinnamon, pepper and tea; the profits then funded further colonial endeavours. Refining the logistics of their operations, colonists established plantations of large-scale monocultures that were tended by enslaved and coerced labour forces, further cementing the relationship between the domination of people and plants. Ecological simplification and destruction of habitats and landscapes proceeded alongside the destruction of the local cultures in far-flung places that were well out of sight from those reaping the benefits in the metropolitan economies.

In the ensuing race for new commodities, plant hunters sent seeds and cuttings back to Britain to test their commercial viability. In the period between 1770 and 1820 Britain had 126 official collectors dispersed across the globe, as well as efficient networks of transport and distribution. Officials of imperial stalwarts the East India Company – effectively an unregulated, outsourced service provider to the government with its own private army – made propitious introductions for plant hunters with local landowners, thus oiling the wheels of access for forays into the inner realms of Asia.

The nascent Kew Gardens and Horticultural Society were both instrumental in purloining plants from pastures afar, commissioning plant hunters to bring back unknown specimens for research purposes, while the burgeoning nursery industry dispatched their own collectors to search for potential new stock. Many introductions were too tender to survive outside in the British climate and needed greenhouse cultivation, but by the late nineteenth century a steady supply of hardy species had boosted sales catalogues and whetted the appetites of the burgeoning middle classes, with their villas and gardens in the ever-expanding suburbs.

Previously, garden ownership had primarily been the province of the wealthy and gardening a closely guarded profession whose practitioners protected horticultural knowledge as an arcane art, but the growth of this new socioeconomic demographic, house proud and with disposable income, witnessed a rise in consumerism orientated towards leisure activities such as gardening. Their march on the outdoor world heralded the birth of the amateur gardener. A plethora of new publications fed their appetite for previously inaccessible horticultural knowledge, such as those produced by prolific garden writer Stanley Hibberd from the 1860s onwards, including the successful *Amateur Gardening* magazine.

This uptake of interest, which led to such a surge in gardening and eventually to the self-proclaimed 'nation of gardeners', was perhaps unsurprising. The maritime, temperate climate of the British Isles, with distinct seasons of warm, wet summers and cool, wet winters, lends itself favourably to growing a wide range of plants. The local flora that developed in Britain after the last glacial period, at the end of the Pleistocene 12,000 years ago, was initially resurrected from seed banks uncovered by retreating ice, whose deposits dated from when Britain was part of the greater landmass of the European continent, or were later introduced without human intervention. Britain's subsequent relative isolation meant that the native flora was limited to around 1,400 species, 47 of which were endemic (exclusive to it). These species are often given short shrift by gardeners or even ostracized as weeds. Their often small but subtle flowers are unable to compete with more blowsy blooms, and their early flowering season fails to endear gardeners longing for displays extending into the enclosing months of autumn. Consequently, the opportunities that newly arrived plants offered gardeners were significant in increasing not only the scope of plants that could be used in a garden, but also a greater diversity of form and extended periods of interest throughout the year.

With this new-found abundance of planting, flower gardens truly came into their own. The Victorian mania for 'carpet bedding' schemes in both private gardens and public parks, named for their resemblance to patterns on rugs, reached its pinnacle towards the end of the century with extravagant displays. This was an industrial-scale objectification of plants in order to create over-the-top displays of pattern and colour from geometric arrays of annual, biennial and tender perennial plants, such as geranium, calceolaria, lobelia and feverfew. The mass-produced and short-lived nature of the plants, which necessitated constant attention and replenishing as they tired, epitomized a highly unsustainable approach to planting. The legacy of the fashion can still be seen today in garden centres and supermarkets piled high with crowded shelves of industrially produced plants at the first outset of spring.

The trend caught the ire of an idiosyncratic Irish exile in London by the name of William Robinson, a garden writer, horticultural influencer and provocateur, who mounted an all-out attack on the practice, criticizing the penchant for pristine plants in perfect bloom needing continuous refreshing and renewal. Cannily grasping the labour and finances needed to support these displays at a time when horticultural skills were limited and greenhouses costly to heat, he fired a calculated broadside attack on these chromatic assaults, which he described as 'a blaze of "colour" which could be almost equally well spread out by the cotton-printer, and which render our gardens generally the most uninviting places in the world'.[18] His disdain for their visual effrontery was equally matched by one for the age of mechanization, something he shared with fellow travellers in the contemporaneous Arts and Crafts movement. It was an inclination that expressed a romanticized vision of an idealized past in tune with nature: 'Today the ever-growing city, pushing its hard face over the once-beautiful land, should make us wish more and more to keep such beauty of the earth as may be still possible to us.'[19]

Bedding wasn't his only horticultural bugbear. He also poured scorn upon the architectural formalism of hedging and topiary as advocated in two books written by architects – John Dando Sedding's *Garden-Craft, Old and New* (1891) and Reginald Blomfield and F. Inigo Thomas's *The Formal Garden in England* (1892) – initiating a heated debate in which he insisted that gardens should grow out of experienced gardeners' knowledge rather than be the product of architects' dreams of idealized plans and patterns of geometry. His aversion to artifice also extended to the gardenesque notion that 'a garden is a work of art, and therefore you must not attempt the imitation of nature in it.'[20]

In opposition to these styles Robinson proposed a looser, more naturalistic planting method, inspired by natural environments and informed by his early career stint at the Royal Botanic Society's garden in Regent's Park, where he tended native plants and collected them from the countryside. His notion of the 'wild garden' was articulated through his weekly publications *The Garden* and *Gardening Illustrated* and most notably his books, *The Wild Garden* (1870) and *The English Flower Garden* (1883).

His naturalistic view was a far cry from the idealized and ecologically simplified version of the countryside advocated by the English landscape movement. Instead, the wild garden would feature 'plants of other countries, as hardy as our hardiest wild flowers', growing as if they were wild themselves, 'without further care or cost'. Drawing upon the influx of newly introduced plants, he advocated long-term planting schemes composed of a mixture of these with natives. The main importance for Robinson was that the plantings were permanent and that plants should be chosen to be appropriate for their sites by 'placing plants where they will thrive without further care'. His intention was that introduced plants would naturalize given the right situations and conditions, making the initial human intervention undetectable (Plate 2).

Robinson elaborated a number of reasons for adopting his planting strategy: hardy plants will thrive and look much better in rough places than in formal beds; they will not look untidy after blooming when planted with other plants in successional flowering; it offers possibilities to grow a greater range of plants; and 'there can be few more agreeable phases of communion with nature.'[21] Key to success was paying attention to the planting conditions. In an astute ecological suggestion which would echo loudly in the garden world a century later, Robinson stated, 'As to soil, etc., the best way is to avoid the trouble of preparing it except for specially interesting plants. The great point is to adapt the plant to the soil – in peaty places to place plants that thrive in peat, in clay soils those that thrive in clays, and so on.'[22] Some of his suggested plant associations trialled at his garden at Gravetye Manor in Sussex, such as double peonies in grass, clematis in a yew tree and tiger lilies in a wild garden, were rather bizarre, but one enduring legacy can still be seen today in the naturalizing of narcissus and daffodil bulbs in grass.

The plant lists in the book look strikingly similar to the catalogues of current-day nurseries, featuring such prairie favourites as aster, asclepias, eupatorium, helenium, helianthus, rudbeckia, solidago and vernonia. These were newly introduced at the time and Robinson was pioneering in their use, especially given that many were plants with weedy vigorous-growth characteristics needing space and therefore unsuitable for formal-style gardening. He suggested the merits of 'numerous exotic plants of which the individual flowers may not be so striking, but which, grown in groups or colonies, and seen at some little distance off, afford beautiful aspects of vegetation',[23] prefiguring the penchant for massed planting in the first decade of the twenty-first century.

Yet despite his perceived attention to naturalism, Robinson was under no illusion that gardens were not a human construct: 'Reproducing uncultivated Nature is no part of good gardening, as the whole reason of a flower garden is that it is a home for cultivated Nature. It is the special charm of the garden that we may have beautiful natural objects in their living beauty in it, but we cannot do this without care and culture to begin with!'[24] And while he railed against the idea that gardens could never be subservient to art, he nonetheless saw an art within gardening: 'Our gardens are beautiful in proportion to their truth to nature. Natural and artistic gardening (synonymous terms) mean the art of expression of the beauty of the vegetable kingdom in gardens.'[25]

Despite all his rabble-rousing, Robinson still retained some conservative inclinations, favouring a degree of relaxed formality around the house featuring herbaceous beds, preferring the wild garden to be kept at a slight remove in less manicured parts of the property such as woodland edges. He was cannily prescient in considering the changing nature of gardening and gardeners, and his ideas about mixing appropriately hardy, introduced plants with natives in a local context show the beginnings of an early understanding in British planting of the ecological requirements of plants which is resonant today.

THE COLOURISTS

While Robinson is often considered the forefather of the classic English garden, Gertrude Jekyll is undoubtably the doyenne whose influence is still very much in evidence in the garden world. As a close associate of Robinson's and contributor to *The Garden*, she pushed the envelope of his ideas, refining the practice of mixed planting and significantly shifting its focus to favour the floral. While well aware of the importance of foliage, form and texture in introducing rhythm and repetition, her emphasis on flowers became the defining feature of her approach. Her most detailed explorations on form, foliage and flower appeared in her book *Colour in the Flower Garden* (1908).

The success of her technique involved orchestrating bountiful blooms to provide interest over as much of the year as possible. The classic Jekyllian border consisted of long, informally planted flower beds, usually framing a lawn path and filled with a mix of bulbs, annuals, perennials and shrubs. Plants were arranged in specific borders to perform at certain times of the year, flowering for a few months each in succession (Plate 3).

Her effective mash-up of the herbaceous border and cottage garden planting styles fused methods from divergent ends of the social spectrum. Reigniting the flame that had been lit by the development of the herbaceous border in the early nineteenth century, Jekyll repurposed it for the new age and the bountiful opportunities offered by an expanded plant palette. Borrowing elements from the gardens surrounding 'the lowly cottage dwellings of the labouring folk' was very much a form of cultural appropriation, given the origins of the style as a practical solution to providing for the needs of manual workers.[1] Her adoption of it as something purely aesthetic stripped it of the social connotations of the working class, repackaging it into a more palatable form suitable for the privileged echelons of society.

The training from her earlier occupation as an artist influenced her view of garden design as a creative practice concerned with the more decorative aspects of plant arrangement. Plants were used as elements to construct an overall composition, as if they were paint on a canvas: 'In practice it is to place every plant or group of plants with such thoughtful care and definite intention that they shall form a part of a harmonious whole, and that successive portions, or in some cases even single details, shall show a series of pictures.'[2]

An infatuation with colour came to dominate her approach, with planting combinations structured around the effects that certain flowers would have when placed together. While colour had its heyday in bedding schemes, she was at pains to protest her distance from them even more vehemently than Robinson had: 'It wants no imagination; the comprehension of it is within the range of the most limited understanding; indeed its prevalence for some twenty years or more must have had a deteriorating influence on the whole class of private gardeners, presenting to them an ideal so easy of attainment and so cheap of mental effort.'[3] Instead, underscoring her class prejudices, she extolled the 'proper' use of colour as an art of sophistication that could only truly be understood by those educated enough to appreciate its value.

Very much influenced by the newly developing science of colour theory and the relationships based around the colour wheel – a device for analysing the effects of associated colours – Jekyll sought out the subtleties of complementary and contrasting tones in their many and varied permutations. Consciously constructed drifts, rather than traditional blocks of plants, were deployed to produce movement through the border, with tonal transitions teasing the eye along in predetermined directions. Meticulously graded changes of harmonious hues were rallied to produce sequences best appreciated by either close-up or distant viewing. The desired effect was of floral tapestries balancing light and dark tones, bright and dull shades, and differing tints, creating visually evocative compositions. She also proposed the idea of minimalist monochromatic borders with restricted plant palettes to provide a predominance of a singular colour. Ideally, a series of them could be arranged in sequence, she suggested, to evoke distinct aesthetic and emotional responses. A degree of Robinsonian naturalism seeped into the peripheral woodland setting of her garden at Munstead Wood in Surrey. This was hardly surprising given that not only had he offered advice on its design, but she also shared his general understanding of the appropriate placing of plants to meet their requirements. But the central part of the garden highlighted her attention to artifice and detail, a defining characteristic of her work, particularly evident in the numerous gardens she worked on with architect Edwin Lutyens. Their partnership delivered gardens clearly defined by his simple structural frameworks of hardscape elements overlaid with her contrasting exuberant herbaceous planting.

Spontaneity had no place in her world view, and contrivance was the means of controlling the unwelcome effects of natural processes. The changing nature of plant forms over the growing season produced bothersome effects such as plants dying back at uneven rates, leaving gaps in the border, which were then either ripped out and replaced with fresh specimens, or filled with plants in pots discreetly hidden behind the suitable foliage of the adjacent vegetation. The 'smoke and mirrors' required to uphold these pictures in their pristine states belied their effortless appearance and effectively hid their unsustainable nature.

Not merely confident of the benefits of her approach, she asserted them in a righteously moralistic manner, one that only privilege can bestow: 'It seems to me the duty we owe to our gardens and to our own bettering in our gardens is to use the plants that they shall form beautiful pictures; and that, while delighting our eyes, they should be always training those eyes to a more exalted criticism; to a state of mind and artistic conscience that will not tolerate bad or careless combination or any sort of misuse of plants, but in which it becomes a point of honour to be always striving for the best.'[4]

Jekyll bequeathed not only her horticultural acumen tempered by her artistic vision, but also this indignant attitude fuelled by a class-based instrumental view of the natural world, in which it was imperative to treat plants as objects to be arranged in a very prescriptive manner. Failure to adhere to this was nothing short of bad gardening and bad behaviour, a dictate that has held sway over many gardeners and planting ideas for the past century.

Her work ushered in the age of the colourists and a century of chromatic obsession in the garden world, in which the holistic aspects of plants were an inconvenient truth, only be acknowledged when they assisted in producing the desired efflorescent effects. Jekyll's influence became the mainstay for the upper-class gardening elite and was quickly disseminated by them to a wider audience through the media, both at home and abroad.

News of her work crossed the Atlantic, where it was picked up by writer and designer Louise Beebe Wilder, who disseminated her ideas through her work in residential gardens across the United States in the first few decades of the twentieth century. The title of Wilder's book, *Colour in My Garden* (1918), was a conscious homage to Jekyll, and the content placed her ideas within a North American context, drawing examples from her own garden, Balderbrae in New York. Her confiding tone to readers encouraged them in their artistic pursuits: 'We are haunted by visions of exquisite colours in perfect harmony, and our aim is henceforth to make the garden a place for broad survey as well as for minute scrutiny; to enjoy, not only the individual flower, but to make the most of it in relation to other flowers.'[5] Another acolyte, Louisa Yeomans King, wrote *The Well Considered Garden* (1915), which came with a preface from none other than her British mentor.

Both women were associated with the burgeoning regional garden club movement, which consolidated its popularity with the formation of the Garden Club of America in 1913. The new club lost no time in establishing a Color Chart Committee, steered for the next two decades by King and influential landscape architect Fletcher Steele. At the time, two colour charts were used by gardeners as aides to floral composition. *Répertoire de couleurs* (1905), devised by the Société Française des Chrysanthémistes and Henri Dauthenay, used oil-based chromolithographs for colour comparison, and was based on work by chemist Michel Eugène Chevreul to ensure quality control with fabric dyes. The other one was *Color Standards and Color Nomenclature* (1912), created and self-published by American ornithologist Robert Ridgway, to standardize the names of colours used to describe birds; it featured 1,115 colours illustrated on aniline dye swatches and accompanied by written descriptions. After much careful scrutiny, the latter was eventually decreed by the club as the official guide and became the standard reference for gardeners applying colourist principles, although the seriousness that the club attached to planting by colour led them to propose that, as a back-up, 'Club Members using the chart should keep in mind the possibility of supplementing the Ridgway with further color scales which, in combination, would make a complete chart for garden use.'[6]

Meanwhile, in Britain, exercising a sense of artistry became key to adherents of the new cult of colour, and the most prominent followers in Jekyll's footsteps adopted both her attention to detail and her careful construction of artifice. Lawrence Johnston arrived at Hidcote in the Cotswolds in 1905 to be faced with little more than empty fields, from which he proceeded, over the next twenty-five years, to fashion a garden. Key to the design was the introduction of formal hedging to tame and frame the space, featuring a mixture of beech, yew, box, holly and hornbeam. This return of architectural planting was a distinct retreat from Robinson's dislike of such devices, and while Johnston toyed with the wild garden in the valleys at the property's periphery, he accepted formality as a means to create a structural backdrop for herbaceous displays.

Johnson's design for Hidcote, a series of geometrically arranged 'garden rooms' laid out around a symmetrical axis, bore the unmistakable influence of garden designer Thomas Mawson's 1900 book *The Art and Craft of Garden Making*, which advised that 'the arrangement should suggest a series of apartments rather than a panorama which can be grasped in one view.'[7] The idea of rooms reinforced the domestication of outdoor space, highlighting the orderly aspects of controlling plants as if they were ornaments within the house, with the hedging clearly demarcating the divide between the interior and exterior, the wild and the cultivated.

Picking up on Jekyll's idea of a series of colour-coded gardens, Johnston filled each space thematically. The Red Borders, the White Garden, the Winter Border, the Maple Garden, the Fuchsia Garden and the Hydrangea Corner were some of the horticulturally defined spaces, each given a clear character by the specific plants and colours within them (Plate 4).

This approach belied an idiosyncratic taxonomic system, categorizing plants by specific features rather than any naturalistic or environmental relationships. A keen plantsman and occasional plant hunter, his interest lay in unusual plants and the unique ways that he could combine them. As an American born in Paris, Johnston's bountiful brashness was largely free of the fetters of British good taste, employing a looser weave than Jekyll, discarding her careful associations and drifts, for bigger and bolder effects; he placed emphasis upon contrasts not only of flower colour but also of texture, shape, size and foliage.

Suitably impressed by his talent, garden designer and writer Vita Sackville-West saw in him a kindred spirit. Her garden at Sissinghurst in Kent, made with her husband, Harold Nicolson, over a period of thirty years from 1930, shared features similar to those at Hidcote, including evergreen hedging and brick walls used to create compartments for planting.

Sackville-West's obvious pleasure in artifice challenged Jekyll's crown as artist gardener with a painterly approach in which flower colour leads the way: 'One has the illusion of being an artist painting a picture – putting in a dash of colour here, taking out another dash of colour there, until the whole composition is to one's liking and at least one knows exactly what effect will be produced twelve months hence.'[8] Her nonchalant tone is indicative of an anthropocentric certainty of getting natural processes to conform to her design intentions.

As with Jekyll, colour was paramount and like Johnston she put into practice the idea of monochromatic planting, including what has come to be Sissinghurst's most renowned feature, the White Garden. 'It is amusing to make one-colour gardens . . . if you think that one colour would be monotonous, you can have a two- or even a three-colour, provided the colours are happily married . . . for instance the blues and purples, or the yellows and the bronzes, with their attendant mauves and orange, respectively. Personal taste alone will dictate what you choose.'[9]

While aware of the general needs of plants, her method when choosing which ones to grow together was largely driven by the visual effects they would produce: 'I observe, for instance, a great pink, lacy crinoline of the May-flowering tamarisk . . . What could I plant near it to enhance its colour? It must, of course, be something that will flower at the same time. So I try effects, picking flowers elsewhere, rather in the way that one makes a flower arrangement in the house, sticking them into the ground and then standing back to observe the harmony.'[10]

Vita's role as high priestess of horticultural sophistication was evident in her role as regular columnist for the *Observer* between 1946 and 1961. Here she dispensed a heady cocktail of observations from her own experience, publicizing her personal preferences and prejudices about plants. 'It is quite true that I have no great love of herbaceous borders or for the plants that usually fill them – coarse things with no delicacy or quality about them. I think the only justification for such borders is that they shall be perfectly planned, both in regard to colour and to grouping; perfectly staked; and perfectly weeded.'[11]

Another garden writer whose influence is still an observable presence was Margery Fish, and her garden at East Lambrook Manor in Somerset, made from 1938, showcased her ideas. Her books *We Made a Garden* (1956), *Cottage Garden Flowers* (1961) and *A Flower for Every Day* (1963) somewhat disingenuously presented an everyday sense of rural domesticity, concealing the class-based privilege that made it possible. Taking the idea of the cottage garden, Fish stripped it further of its humble origins yet attempted to retain some of its earthy associations of honest labour and connection to the land, accentuating them in a form of 1950s shabby chic.

Unabashedly romanticizing the past for its assumed authenticity, Fish proclaimed:

> Nowhere in the world is there anything like the English cottage garden. In every village and hamlet in the land there are these little gardens, always gay and never garish, and so obviously loved. There are not so many now, alas, as those cottages of cob or brick, with their thatched roofs and tiny crooked windows, are disappearing to make way for council houses and modern bungalows, but the flowers remain, flowers that have come to be known as 'cottage flowers' because of their simple, steadfast qualities.[12]

Her eager insistence on the need for gardens always to be in flower reflects the enduring mindset that plants are there to perform for people, and that seasonal change, rather than being something to celebrate as plants' responses to environmental factors, instead should be mitigated by corralling as many plants together as possible. No natural habitat displays continual blooming, yet the arrogance of artifice employed by gardeners attempts to deny ecological reality and supersede it by careful calculation: 'A garden that is to be good always needs careful planning. To get flowers for every day in the year means that no space must be wasted and the plants chosen must have flowering seasons to cover the whole year.'[13]

While Fish's floral fetishism applied a rosy lens to the class origins of cottage gardens, they were less of a concern for another garden writer, Rosemary Verey, who unashamedly assimilated the style, mixing it with traditional herbaceous borders and architectural topiary at her garden at Barnsley House in the Cotswolds. Less grand than Sissinghurst or Hidcote – but still decidedly upmarket – the garden, designed from 1951, played off structural evergreen forms of clipped shrubs, knot gardens and box-edged beds against self-seeded geraniums, wild strawberries and forget-me-nots that littered pathways in calculated acts of insouciant disobedience.

An impressively architectural laburnum arch, underplanted with seasonal bulbs, not only provided a visual framing device across the garden, but also created maximum impact in terms of colour. Verey's take on the colourist approach was sequential: 'My aim is to make the beds as full of colour as possible right though from March to autumn. They start with mauve, yellow and white – crocus, narcissus and hellebores – then some are taken over with white, yellow and orange tulips, and others with red tulips, all underplanted with forget-me-nots, cowslips and primulas.'[14] Verey's tactic of disguising the overall formality of the garden with wilder touches was central to her method of mixing previous garden styles, as explained in her book *Classic Garden Design: How to Adapt and Recreate Garden Features of the Past* (1984); it quickly became the comfortable default style for gardens in the area, and continues to define them today.

In the less leafy landscape of East Anglia, Victorian island display beds made a dramatic comeback in a mid-century manoeuvre that has endured for seven decades under the eagle-eyed aegis of the Norfolk-based Bloom horticultural dynasty. In 1946 nurseryman Alan Bloom purchased Bressingham Hall and land at Bressingham, where he began developing a large-scale garden in 1953. Rather than using annuals as the Victorians had done, Bloom drew upon the family nursery's stock-in-trade of perennials, utilizing five thousand plants laid out in island beds, set in rolling lawn over a 2.5-hectare/6-acre site, when it opened in 1962.

Despite the massed arrays of plants, Bressingham's design is a thinly disguised take on the gardenesque, with the island beds approachable from all sides, offering a variety of vantage points to appreciate the arrangements. The primary focus on the aesthetic combinations of plants, based upon form and colour, drew on techniques from both the traditional herbaceous border and annual bedding.

The influence of the kaleidoscopic style was pronounced in its popularity throughout the 1960s and 1970s, particularly as the concept was scalable to suburban gardens. The appeal to the amateur gardener was clear; it offered a distinct rebuff to the upper-class country house style, dependent upon large tracts of land and a lot of labour for their upkeep. It also eschewed the looser, less formal, cottage tradition, opting for a more modern sense of order.

While managing the garden at Tintinhull House in Somerset for the National Trust during the 1980s and 1990s, garden writer and designer Penelope Hobhouse struck a note of growing realism about 'appropriate planting', recognizing that various factors limiting the palette of plants to use could actually be a creative constraint: 'Garden colour planning is not only arranging harmonies around existing garden features; in many cases you decide on a definite colour scheme and then choose the plants which will make it live. In fact, much of the basic colour arranging depends on both factors, rearranging plants to make better compositions and finding plants to perfect a colour picture as yet on paper or even still an image in your mind.'[15]

The garden at Tintinhull was on the Hidcote/Sissinghurst model of formal evergreen structure used to create compartments and backdrops for mixed herbaceous planting. Originally laid out by Phyllis Reiss in the Arts and Crafts manner in the 1930s, it bore the unmistakable hallmarks of Jekyll's influence, featuring a gold-and-purple border, a white garden, and a planting of blue flowers and purple foliage based on shared bluish pigments in each.

Working in the garden, Hobhouse implemented colour theories in practice using both flowers and foliage. Her book *Colour in Your Garden* (1985) was the most thorough exploration of the subject. Delving deep into the 'mystery' of colour, investigating its science and psychology, she strove to apply dimensions such as hue, value and intensity to create simultaneous contrasts and associations in a more rigorous manner than any of the former colourists. By providing such an authoritative grounding, marrying art and science, she managed to further legitimize the appeal of planting by colour as the primary determinant of good gardening.

The approach to planting with colour reached its pinnacle in the 1990s at Hadspen in Somerset, where Canadians Nori and Sandra Pope transformed an eighteenth-century, parabolically shaped walled garden into an orchestrated array of tones. Eschewing the prevalent practice of contrasts in colour, shape and size in favour of carefully considered tonal arrangements, their approach presented a sophisticated means of deploying colour using gradients and subtle shifts through hues to generate a sense of momentum. Very much inspired by the idea that gardens are an art form, the painterly colour theory that the Popes expounded in *Colour by Design* (1998) likened planting design to the compositional techniques of music, creating contrasts and movement 'from foliage to flower and back and forth . . . very much like a score of music as it unfolds'.[16]

As they rented the garden from the estate owners, the Hobhouse family, a certain pedigree came with the territory. Penelope Hobhouse had previously gardened there but the garden had fallen into neglect after her departure, providing the Popes with a blank-ish slate to work with. Working stringently within the colourist mode, they took direct inspiration from both Jekyll's *Colour Schemes for the Flower Garden* and also Hobhouse's more recent update in *Colour in Your Garden*.

The monochromatic colour-themed borders they created recall the singular approaches of Sissinghurt and Hidcote yet take them a step further, developing the use of colour as a theme, moving through shades of one colour to another to create harmonies. 'Using a single colour or a group of colours, we wanted to refine that, like say use a scarlet and work through all the plants that have this pure scarlet flower and try to work with pure tones.'[17] They saw this sense of progression:

> As a system of creating dramatic tension, increasing perception, and enhancing mood change, we advocate planting in a developing monochrome. Less is more: by using monochrome (single colour) plantings at Hadspen, we can closely control the colour shift, the saturation of colour and the tonal change from dark to light. Using a single colour also makes it possible to focus on foliage and on flower shapes, on the rhythm and structure, and of course, on the full impact of what the colour offers emotionally.[18]

Colour also dictated plant placement according to the optical and psychological effects produced by certain colours at varying distances and when moving from light to dark, pale to saturated. Their work took planting by colour to new heady heights, while recognizing the immense input involved in achieving this ambitious end of horticultural artifice. The importance of seasonal succession to deliver eye-popping chromatic displays for seven months was a task that entailed intensive schedules of planning, cutting back, replacing plants and pinching out flower buds to delay blooming.

MODERNISM:
FORM, FUNCTION, FOLIAGE

As an architectural style, modernism entered the early twentieth century with grand designs and socially progressive ideas, revolting against the decorative stylings then dominant in the arts by proposing instead an aesthetic more in tune with the times and reflecting the increasing industrialization of urban society. The ideal of creating buildings as 'machines for living' marked a move towards clean lines and minimal open spaces, discarding the ornamentation of the past. Instead, everything in the design had a purpose expressed through the amalgamation of function, form and materials. Modernism established a foothold in the landscape in the United States in the 1930s through a group of landscape architects whose work came not just to define a style but also to reflect the development of American society in the mid-twentieth century.

The radical break with the prevalent Beaux-Arts style of landscape design, premised upon European Renaissance and Baroque formality and geometry, was vociferously articulated by three students at Harvard University determined to chart new territory. Garrett Eckbo, Dan Kiley and James Rose shunned the design direction being foisted on them by their tutors and began to develop a new language of landscape based upon an integrated view of form and function in which the spatial design of the garden was inextricable from its use. In particular, they addressed the developing lifestyle preference for integration of indoor and outdoor space, viewing house and garden as a unified spatial unit, and the integration of architecture and landscape as an opportunity to ensure the integration of people with their environment.

In their vision gardens were no longer empty spaces waiting to be filled with plants. Rose was derisive about the penchant for gardens as pictures painted with plants: 'We cannot live in pictures, and therefore a landscape designed as a series of pictures robs us of an opportunity to use that area for animated living.'[1] Even more dismissively: 'Ornamentation with plants in landscape design to create "pictures" or picturesque effect means what ornamentation has always meant: the fate call of an outworn system of aesthetics. It has always been the closing chapter of art which had nothing more to say.'[2] In his use of plants, aesthetics and pragmatic concerns were indivisible: 'The intrinsic beauty and meaning of a landscape design come from the organic relationship between materials and the division of space in volume to express and satisfy the use for which it is intended.'[3] The inherent qualities of plants were never merely visual but also defined form and directed use: 'All plants have definite potentialities and each plant has an inherent quality that will inevitably express itself . . . Plants are not applied to a preconceived ground pattern, they dictate form as surely as do use and circulation. Their character develops from the actualities that have been solved rather than obscured.'[4]

Rose was based in New York and Kiley in New Hampshire and Vermont, but it was in the conducive climate of California that the spatial synthesis of interior and exterior really hit a high. Working primarily in the state, Eckbo was able to employ the weather as another element in his work, where outdoor living was gaining purchase among the growing demographic of homeowners. He recognized that the democratic aspirations of the new consumer culture longed for the original ideals of gardens, but new forms were needed that were relevant to the times: 'As social inequities become more complex, those who have more than the average, and more than they need, tend to express or flaunt such surpluses in the scale, spreads, character, and quality of their homes, in both architecture and landscape surroundings. Paradise becomes private, though occasionally opened to the masses on holidays. It becomes apparent that it requires substantial land, materials, skill, and manpower to reproduce paradise so long after the Fall or Exodus. For the common man, dish gardens, patios, or suburban backyards may provide symbols or memories of the Paradise of the rich.'[5]

His vision was a socially focused and holistic one of people, plants and place. He worked with many modernist architects and found this provided the ideal partnership. He understood that people and their behaviour were part of nature, and that the house, garden and wider landscape beyond were all just part of the continuum of the environment. Eckbo believed that it was the designer's role to ensure that these were all integrated and worked together.

As many new materials developed during the war effort in the 1940s began to make their way into consumer society, the modernist designers were keen to embrace them within their brave new vision. But despite the novelty these new hardscape materials offered, plants provided a unique sense of freedom for the designer: 'Planting design is the one kind of design which has few boundaries derived from technical or site necessities.'[6] And plants played an important role in shaping the spaces: 'The object of planting is not only to work into the garden scheme all the different kinds of plants which you would like to have for one reason or another. It is also to put them together in a continuous framework or pattern which will unify the garden space somewhat as the structure of the house unifies the space within it, although more loosely.'[7]

Plants were considered in their entirety for their size and shape, textures and colours, but reducing the palette ensured the visual and functional integrity of the garden. Eckbo believed that the designer's 'objective is the integration of continuity and entity, by the integration of planting design and structural design in the larger art of spatial design . . . Every plant, no matter how low, how prostrate, how massive, matted, or solidly bushy, how fastigiate or billowing, is nevertheless a construction in space and an enclosure in space.'[8] The layering of space through plants was crucial. Grasses and ground cover supplied the lower level, shrubs and herbaceous the middle, and trees the upper, providing shelter and enclosure. The spatial arrangement of these levels in different areas of the garden was unified within the overall garden layout to create a three-dimensional unity. Eckbo's enthusiasm for the creative opportunities this provides was evident:

'Plant material as an aggregation of units of unlimited variety in form, size, colour and texture, has aesthetic possibilities which have scarcely been scratched.'[9]

Trees in particular played an important role, as did structural evergreen shrubs and perennials, which had an all-year-round architectural bias denied to many herbaceous plants. Colour was not simply framed around flowers, but applied to all aspects of the plant, with the considerable range of foliage and structure providing definition of the plants themselves as well as the space as a whole.

Eckbo had a keen ecological understanding of site conditions and the function of plants, and believed that an amalgamation of ecological, botanical and horticultural knowledge was essential for plant selection. He also cannily picked up on the growing movement towards garden owners wishing to spend less time on upkeep of their gardens, something he addressed by sensitive plant selection of hardy plants that were happy without attention and interference. 'Planting must be worked out in terms of selection, arrangement, and maintenance. These are all three mutually interdependent and retroactive.'[10] This complete vision, in which the ways gardens were defined spatially and visually were inextricable from how people used them, was articulated in his books *Landscape for Living* (1950) and *The Art of Home Landscaping* (1956), which democratically conveyed his ethos to professional designers and homeowners alike. His later work *The Landscape We See* (1969) addressed the bigger task of how landscape design, as a problem-solving discipline, could face the growing environmental crisis, outlining a mindset that future generations of landscape architects would adopt.

Also on the West Coast, another landscape architect, Thomas Church, had the same concern that gardens should be for people. A decade older than the young upstarts, Church was nevertheless part of the zeitgeist. While his earlier and later works were in a more traditional vein, during the period 1947 to 1952 his work exhibited a distinctly modernist bent. The biomorphic shapes of his designs lent an organic character to spaces, and the swimming pool for his design at the Donnell garden in Sonoma County in 1948 was a definitive statement of the style, merging landscape and lifestyle; it has since become an iconic image. His planting in this period was very much concerned with preserving and enhancing the majesty of the architecture, with accent planting around the house and crisp, controlled hedges: 'By their shape, colour and foliage texture, plants show that they are intended to enhance architecture not hide it.'[11]

Modernism's reach made it to South America, where landscape architect Roberto Burle Marx created a unique and visually distinct body of work that increasingly gained international acclaim. He was unlike other modernist pioneers in that plants were very much at the fore of his practice in a bold and bright manner. While he was undoubtably an adept plantsman with a keen knowledge of local flora, the manner in which he deployed plants in his schemes was premised upon a high-impact aesthetic appeal rather than any ecological notions about communities or interactions. Planting in huge swathes of monocultural blocks with grand patterns based upon flower or foliage colour created a highly distinctive way of shaping space, particularly the ground plane. He was less concerned with the integration of three-dimensional space with materials, like the other modernists; indeed, his work echoed the carpet bedding of the Victorians passed through the filter of abstract art.

The modernist grip never really took a hold in Britain in the same way that it did in the United States. It was always going to be an uphill battle against a more conservative culture and a long-established ornamental garden tradition, but it did make some inroads and a home-grown version even had a lasting effect on British garden culture.

Canadian-born landscape architect Christopher Tunnard was an early adherent to the style. The influence of leading modernists Le Corbusier and Alfred Loos, as well as Japanese gardens, led him to advocate simplicity in design that fused function and aesthetics into one. Rather than drawing inspiration from paintings as the English landscape movement had, or trying to create Jekyllian pretty pictures with flowers substituting for daubs of paint, he instead considered the entire design to be a creative act – the garden a work of art in itself. His rallying call *Gardens in the Modern Landscape* (1938) heralded the arrival of modernism into the field of landscape architecture, predating the work of the Americans and influencing many of them. Scathing of the complacency of design, he complained: 'Contemporary Garden design has not yet even caught up with contemporary trends in architecture. It is to be hoped that in the near future garden-makers will become aware of this fact, and that instead of rehashing old styles to fit new buildings they will create something more expressive of the contemporary spirit and something more worthy of the tradition to which they are the heirs.'[12] Clean, clear lines and spaces free from clutter and ornamentation were key to his approach in attempting to create a harmonious balance between architecture and landscape by letting 'space *flow* by breaking down division between usable areas and incidentally increasing their usability'.[13]

Traditional planting also incurred his wrath: 'It is a subject surrounded by fetishes. How many gardeners can look at a plant nowadays without thinking of its rarity, its sentimental value, or its temperamental ways under cultivation?'[14] Instead, he looked towards plants in themselves for structure and integrity: 'To understand the character of plants and to use those which can supply a formal quality of their own, without insisting on size of bloom or certain fashionable shades of colour, is the aim of the modern garden-maker.'[15]

His approach was an act of high artifice, highlighting plants for their architectural qualities. Being used alone, often in planters sparsely arranged on geometrically paved terraces and courtyards, or laid out in geometric groups allowed their forms and structures to rise to prominence unfettered by any surrounding distractions, the clean lines of the architecture providing a neutral yet complementary backdrop. While clearly drawing on the Japanese tradition of minimal planting, it also recalled the gardenesque's obsession with plant forms and the way they showcased them to be regarded from different vantage points. However, Tunnard's fusion of them into the entire spatial setting, blending form and materials, transcended both previous methods.

Architectural plants with interesting leaf shapes, colours and textures, elegant forms, and fruit rather than flowers were Tunnard's preference. Hardy and tolerant of varying conditions and providing all-year-round interest, they generally required little maintenance. His plant palette included various species of phormium, bamboo, ivy, vinca, viburnum, euphorbia, hosta, fatsia, cotoneaster and tiarella. These plants became the basis of the Institute of Landscape Architects'[16]

Basic Plant List, a standard reference for practitioners, and later reappeared as star features in many gardens of the 1990s.

Pioneering landscape architects Sylvia Crowe and Brenda Colvin were both instrumental in progressing design in the modern world of the postwar years, working on similar types of large-scale infrastructure projects in the rapidly changing industrial landscape, such as power stations and reservoirs. Their work was firmly rooted in their experiences on the smaller scale of garden design in the 1930s. But, though not as radical or rabble-rousing as Tunnard, they nevertheless embodied the same structural sensibility and, like the American modernist designers, sought to reconcile the relationship between people and the landscape. Their work looked backwards and forwards at the same time, addressing the changing social needs of gardens but with a hankering for many aspects of previous garden styles. They articulated their outlook in *Land and Landscape* (1948) by Colvin and *Garden Design* (1950) by Crowe.

Each viewed planting primarily as a functional way of structuring spaces – creating division, providing shelter and mitigating pollution – and they were both keenly aware of the interdependence of form, colour and texture. Plants were viewed as key elements of the whole design; they did not focus on the decorative aspects of individual plants. As Colvin explained, 'Our appreciation of plants depends on their setting and grouping. The relationship of each group to its neighbours, to its background and all its surroundings is of far greater consequence than the individual beauty of any single plant or flower.'[17]

Colvin appreciated the basic principles of ecology in terms of climate, soil and the influences of human and other creatures in the landscape. She understood plants in terms of communities, succession and the symbiotic relationships between animals and organic cycles. She was also forward-thinking in understanding the importance of plants in preventing soil erosion, and in recognizing the ecological threat of declining habitats for wildlife. But in her view plants were primarily to be used for functional structure on a large scale, creating division, accentuating gradients, framing views and complementing architecture in a Reptonian manner: 'Plants, whether we are thinking of trees, grass or flowering herbs, give beauty in the surroundings of human life but they rarely give of their best unless they are serving other fundamental needs as well.'[18]

Colvin protested that the full realization of all aspects of garden design suffered as a result of too much emphasis on masses of colour in planting schemes, a situation that 'has happened to a great extent in England since the introduction of countless hardy flowering plants from all over the world; the interest in good garden design seems to have been in inverse ratio to the taste for cultivating flowers.'[19] She believed the problem could be attributed to unrealistic planning and wishful thinking: 'The amateur tendency is to regard Flower colour as if it was the . . . only thing to aim at in Gdn Design. Hence the poor results. In imagination the flowers are seen permanently – but not in real life.'[20] She proposed to remedy the imbalance between flower colour and other components of design through the employment of a carefully calculated proportion of foliage to flower of 7:1.

Recognizing that plant groups were key in forming the visual character of landscapes, she advocated a minimal selection, believing that consciously designed 'groups of single species usually give the best results, and although in some cases this may limit the number of species used, it makes for more enjoyable living conditions than a mere collection of horticultural specimens of wider variety.'[21] Their arrangement should 'engender a sense of anticipation and interest by the progression from one interesting plant group to the next in a rhythm, giving definite contrasts without loss of unity'.[22]

In true Robinsonian fashion, Colvin embraced the amalgam of native and introduced species, such that 'Any plant which can propagate itself, hold its own, and take its place as a member of a natural plant association is for the purpose of the landscape designer the equal of one which is truly indigenous.'[23] But in defiance of the ethos of the wild garden and in an unashamed embrace of artifice, she wrote: 'It is not suggested that ecological considerations need overrule other principles of planting design, nor that we should apply them to all planting schemes. There are many sites, such as parks and gardens in towns or related to human habitation of any sort, where purely visual and aesthetic standards unhampered by ecology are appropriate.'[24] To this end she proposed grouping together plants from different habitats and parts of the globe, acclimatized to different types of soil and degrees of moisture, although this was to be aided by modification of the soil in specific areas, where necessary creating raised beds and providing dedicated irrigation in order for them to survive. Such assistance ensured that 'it makes possible the use of almost any kind of foliage in any position where it may be required.'[25] She advocated using plants structurally and texturally, in particular evergreens and those with distinctly architectural appeal such as acanthus, rodgersia, phormium and fatsia.

These plants played an aesthetic role and a functional one in addressing the lifestyle needs of garden owners: 'A system of planting which shall eliminate much of the repetitive tasks, leaving scope for those more creative jobs we all enjoy . . . Permanent ground cover planting is one of the directions of this research . . . For the designer, this use of plants as the accompaniment or secondary theme, looking after the ground around primary groups, becomes an element of beauty and promise.'[26]

Crowe shared Colvin's planting pragmatism, describing the functional aspects of designing with plants in order to provide shelter, seclusion, shade and spatial division, articulated through architectural form and accentuated by textural contrasts. She was more inclined than her colleague to accept traces of tradition into her work, acknowledging that while the importance of planting was structural, nonetheless, 'Planting may be either part of the structure of a garden or its interior decoration.'[27] Her reverence for Hidcote's fusion of formality and herbaceous trappings was seemingly at odds with any pure modernist tendencies she may have harboured. Her acceptance of the mixed herbaceous border was also a concession to the persuasive power of the English garden tradition, although she was aware of the pitfalls of planting by colour when it 'is often considered only in relation to the plant's flower. Yet the general tone of the plant and its appearance when out of flower is quite as important, or even more so, since the flowering period may be short. The most carefully chosen flower colour will not compensate for leaves which are out of key with their surroundings.'[28] Instead, she advocated following the lead of the way that colour is displayed in natural landscapes, but translating it into a garden setting rather than slavishly adhering in a form of mimicry. Despite such advice, she devoted much of the section on herbaceous plants to describing how to harmonize colour combinations of specific plants, as well as enthusing on the appeal of monochromatic borders.

Like the other modernists, she was keen to consider plants in a more holistic manner:

> These attributes taken together – stature, form, pattern, texture and colour – add up to the plant's character; that indefinable personality which decides whether plants will look right together, and whether they will fill the particular place in the design for which they are intended. The best associations are between plants which have one element in common and another contrasted. Complete contrast in all elements can be used for special emphasis, but repeated too often the effect is restless, lacking the unity given by a connecting link of similarity.[29]

Despite Colvin and Crowe's modernist sensibilities, their planting was really just an accumulation and expansion on previous planting styles, consolidating the functional and aesthetic aspects within the traditional context. But the social sea change that modernism was responding to was ushering in a new public conception of gardens, as was observable in the June 1960 issue of popular British DIY magazine *Homemaker.* Attesting to the growing popularity of the visions proposed by Eckbo and Tunnard, its cover announced the arrival of 'The Outdoor Room' and provided readers with the lowdown on how to design it, buy the furniture and lay the floor. The editorial content, explaining the importance of creating a link between house and garden, marked the idea of exterior space as a lifestyle necessity seeping into the British cultural consciousness. As the article made clear, 'With the ending of spring comes the vision of balmy days, deck chairs, iced drinks, and lazy luxury under the blazing sun. But isn't it depressing to look from your French windows and find a wilderness? Don't despair. With a little outlay and a lot of energy you can have a smart living area out of doors and a luxury setting for your siesta, and at the same time add considerably to the value of your property.'[30] This amateur approach obviously eschewed the services of the designer, but also had little truck with plants, suggesting that they could in a very minimal fashion be conveniently arranged in pots and planters, or arrayed on shelves to enliven walls.

It was a former employee of both Colvin and Crowe who picked up on this zeitgeist, and really grasped the element of modernism that related to social changes in Britain. The time John Brookes spent in the offices of Colvin and Crowe placed him in the midst of an exciting period of new possibilities from the crossover between architecture, landscape and art. The influence of the Californian modernists struck a note with him – his interest was in democratizing gardens for the wider public rather than simply for the wealthy. Also, increasing prosperity and an optimistic sense of a new world dawning that was evident in 1951's Festival of Britain laid the foundations for his thinking and honed his focus towards a minimalist approach to creating simple, maintainable gardens.

Brookes managed to successfully articulate a design style that responded to the age of rapidly increasing consumerism focused upon lifestyle and social mobility and a public with an appetite for gardens. The spread of suburbia and homeownership produced a new demographic with smaller gardens and disposable income to spend on them. Brookes's approach to elevating these traditionally undervalued spaces was a clear rejection of the conservative tradition of the country house style with large herbaceous borders or the nostalgic rural cottage garden. He recognized that 'A garden is essentially a place for use by people, not a static picture created by plants; plants provide the props, the colour and texture, but the garden is the stage and its design should be determined by the uses it is intended to fulfill.'[31]

He developed a signature style relating gardens to the architecture of the house, playing with geometry, based upon a modular grid approach, in order to ensure an appropriate sense of scale and proportion. He initially outlined his approach in his 1969 book *Room Outside*, and followed it with a stream of publications over the next few decades expanding on the central tenets of his methodology.

Echoing Tunnard, he was clear that 'Modernist designers sought out plants of form and shape that worked visually with their designs, did not smother them . . . Planting as such was not important – individual sculptures were.'[32] Brookes's method of plant selection was firmly rooted in this tradition, and while plants were structurally important, they were not the guiding lights in his designs: 'It is the set and its conditions, the designing and planning of the project, then its structure and manner, and the style of the house and client, that guides me on my subsequent plant selection, and this comes last in the process. To be more precise: I choose plants that add to my composition based on the overall structure of the plant and/or its leaf, its texture, possibly its stem, whether deciduous or evergreen, and only lastly its flower colour, which is transitory.'[33]

This view of planting as a decorative part of the design vision, subservient to the various constituent parts of the overall lifestyle package, marked a change in the form of cultural capital invested in gardens. Breaking from the dominant expression of taste expressed through the sophisticated class-based artistry of colour-orientated floral planting, it aided a gradual regression of plants in gardens from foreground to background.

Contemporary garden design was forged in this crucible of social change, which has seen an increasing paucity of time available to devote to working with plants in gardens. Low maintenance has become the modern mantra. People want to enjoy their exterior spaces without having to toil to keep them looking good, and this has only been achievable through a reduction and simplification of planting within gardens.

The legacy of *Room Outside* has shaped the garden world as much as the influence of Robinson and Jekyll did. The idea of the garden as an extension of the house, and an essential living space, abounds in magazines, makeover programmes and adverts. With space at a premium, gardens are real estate and plants are seen as lifestyle dressing.

In accommodating this trend, contemporary garden designers are to an extent diminishing the rationale that makes the profession unique: the fact that plants are its core material. This growing tendency has seen a rapid rise in outdoor kitchens (usually only a few metres from the indoor kitchen), fire pits, heaters, dining tables and all sundry accoutrements, which encroach on outdoor space from indoors. While there is nothing wrong with encouraging more people to spend more time in their outside spaces, the fact that these are becoming less plant-filled mini-paradises and more rather sterile rooms, with the attendant housekeeping required to keep them free from other irritating organic life forms, is a great missed opportunity for increasing urban biodiversity and its attendant benefits.

THE PINNACLE OF PICTORIAL PLANTING AND THE PATH TO BIODIVERSITY

If any gardener could lay claim to the high office of extravagantly colourful artifice, it must surely have been self-styled garden provocateur Christopher Lloyd, whose work at Great Dixter in East Sussex animated the garden world with equal amounts of admiration and ire.

The garden framework, laid out by his father, Nathaniel, starting in 1912, featured a similar spatial strategy to Hidcote and Sissinghurst, with a series of garden rooms divided by yew hedging and topiary. Within each of these Christopher created extravagant arrays, in his idiosyncratic style, of various perennials, annuals, bulbs and shrubs: 'The borders are mixed, not herbaceous. I see no point in segregating plants of differing habit or habits.'[1]

His attitude that rules were for breaking was manifested in what often appeared to be a garden of clashing contrasts, much to the annoyance of those with more traditional horticultural inclinations: 'Every generation has to make its mark in its own way. I think that copying the past is a cop out.'[2] His impish iconoclasm was fuelled by a contrarian disdain for conventional good taste and a lifelong enthusiasm for plants: 'One of the great joys in gardening is to be carried away by a new plant, the sight of it and the feeling it could be yours.'[3] With his seemingly devil-may-care attitude to planting in bold displays of horticultural shock and awe, he disregarded convention by placing tall plants at the front of the border, previously considered inappropriate. As a prolific writer, Lloyd's numerous books often contained the phrase 'for Adventurous Gardeners' in the title, encouraging the readership to accompany him on a journey of novelty and risk-taking beyond the bounds of horticultural good taste.

Yet for all the flamboyant pick 'n' mix chaotic effect of the garden, it was a construction of careful consideration. Colour was certainly an important element of the design, and writ large in the garden, and unlike some of the more tastefully restrained colourists, he was unafraid to use the spectrum: 'Given the right circumstances, I believe that every colour can be successfully used with any other . . . I have no segregated colour schemes. In fact, I take it as a challenge to combine every sort of colour effectively. I have a constant awareness of colour and of what I am doing.'[4]

For Lloyd planting design involved a good deal more than simply flower colour. Given the vast arsenal of plants that he employed, all their various qualities had roles to play in creating the bigger picture on display: 'Heights, shapes and textures, as well as season of comeliness, are all factors to be considered.'[5] Appreciating the whole of the plant meant that 'foliage is more important than the flower, because the foliage is there for so much longer, it's got so much structure, you can get so much more size and fabric into a leaf than you can into the majority of flowers, so I would say that the flowers . . . are a bonus and the foliage is what you look at first.'[6] Disregarding the Jekyllian strategy of planning borders for timed climaxes in particular months, Lloyd instead preferred to achieve a sustained crescendo of colour, form and texture (Plate 5).

His attitude towards specific plants was based upon his experience of using them, and those deemed worthy of remaining in the garden were those that grew well and whose shapes worked together in combination with each other, while those that fell foul of his favour were given short shrift, as if he were a head teacher dismissing students. 'Plants that are grown close to one another need to be able to help each other, visually. For instance, one that creates a haze of small, variegated leaves needs either to have an interesting structure as a plant, by way of compensation . . . or to have neighbours or a backdrop with more sombre, perhaps bolder and contrasting foliage, rather than other small-leaved, variegated plants, which, carried to extremes, will produce the chaotic dog's-dinner effect.'[7]

The rationale of his approach was premised upon continuous change, something only possible with labour-intensive management and the costs associated with it. The class privilege that affords this also underscored his outlook and ethos: 'Dixter's is a high maintenance garden; I make no bones about that. It is effort that brings rewards. There are many borders and much work goes into them. Labour saving ground cover is not for me. If you see ground cover, it's there because, first and foremost, I like it. If it does also save labour, that is an incidental benefit.'[8]

The high entertainment value of Great Dixter is provided by constant change as plants develop naturally with the seasons as part of their life cycles, but also through artificial intervention to up the ante and accentuate such processes. In a manner similar to Jekyll's filling in gaps after flowering, succession involves a continual process of renewal, with plants regularly replaced and bedding plants changed as often as three times a year: 'That's the great thing about bedding, you can change it as often as you like, or you can repeat it; if something goes wrong you can forget about it.'[9] Praising the benefits of the complexity of the method, head gardener Fergus Garrett suggested, 'It's a high-maintenance garden and, as I think somebody said, "low maintenance means low braintenance", and we believe that. It's high input, high output.'[10] While acknowledging that such an approach may not be for every gardener, he adds, 'If you're not interested, obviously you'll want it to be low maintenance, but then you don't get much of a thrill from what you have done. And I think that every plant and every scene should give you a bit of a buzz.'[11]

While Lloyd's methods took the traditional horticultural approach of English gardens to its zenith, a contrasting approach was also evident in his outlook, as evidenced by the meadows on the edges of the garden, which exhibited a decidedly more ecological sensibility. Lloyd used the meadows as a visual and contextual contrast to the rich herbaceous planting in the other areas of the garden, and also as an acknowledgement of the important historical typology they represent, as a managed form of planting on a different scale to the domestic and one that introduced a wilder sensibility. While the meadows were started by his mother in the 1920s, Lloyd extended the idea to the front entrance of the house, which garnered criticism from traditionalists in the garden world, who believed meadows were not something appropriate for gardeners. But it signified a growing understanding of a changing climate and of the different and important ways in which plants interact when not assaulted with constant intervention, and the important benefits this brings for biodiversity: 'As a conservationist concept, meadows take us back to nature and away from the gaudy trappings of civilization.'[12]

More recently, the appreciation of these organic processes has also fed into parts of the garden itself, as Great Dixter has continued to flourish under the watchful eye of head gardener Fergus Garrett since Lloyd's death in 2006. Garrett has been instrumental in introducing new ideas, with more ecologically orientated intentions, to the traditional framework of the garden. A biodiversity audit of the property has revealed the species-rich nature of garden ecology at the site, benefiting from not simply the floral garden areas but the meadows and surrounding rural area as well.

Garrett's approach has involved creating areas within the garden that display a greater complexity, which allows for seasonal change without necessitating the effort of replacing plants, but rather involves a process in which there is 'one plant taking over from another to fill those gaps, so you've got those multilayered borders, mimicking what happens in the wild but just using ornamental plants. These areas more or less look after themselves.'[13] Plant selection and knowledge of their characteristics are key:

> What you are doing is picking the right plants to give you this multilayered community that comes up one after the other. And key to this is picking the right plant . . . you'll use plants that are suited to your conditions, that can grow for you, that you like . . . it mimics what happens in woodlands, the wood anemones are thinking there is a canopy of oak trees over the top of them, or it mimics what happens in meadows because early flowering stuff is then overwhelmed by grass or other vegetation that takes over in a prairie, one lot gives way to another . . . the clever bit is finding the plants that are in harmony with each other.[14]

RIGHT PLANT, RIGHT PLACE

After Robinson's observations on naturalizing plants, an awareness of the importance of placing plants according to their cultural requirements became increasingly incorporated into common practice. The most obvious advantage was to ensure that plants didn't die, but beyond this, the colour-coded displays of the British tradition had little interest in grasping the underlying logic of what plants really needed, or of knowing the science that actually explained it all. The fortune of clement temperate conditions in the Cotswolds and the south of England generally meant that a wide range of plants found these areas conducive to the ecological ranges they were able to survive in. It wasn't until gardening against the odds in the driest region in Britain that a local gardener put plants into a wider perspective, and the right plant in the right place.

As a native of East Anglia, nurserywoman Beth Chatto had never experienced the generous and forgiving conditions that some other gardeners had enjoyed. Instead, she struggled with plants that refused to perform as they were generally expected to: 'Because I cannot grow many of the plants that gardeners living in wetter parts of the British Isles grow so easily and well, I have to go back to the origins of the plants to find those that naturally need a low rainfall.'[1] A visit to the Swiss Alps, where she saw many familiar garden plants flourishing, provided a Eureka moment:

> And here I was seeing for the first time how right plants look when they grow together, in association, in the conditions to which they have become adapted. I noticed too that the associations varied, whether we went a thousand feet or so higher, or whether we were on the north side of the mountain where the conditions were cooler and shadier. I learnt that some plants, although they appeared to be growing in poor gritty soil in blistering bright sunlight, were in fact revelling in the cool trickle of melting snow water that ran not far below the surface.[2]

These groupings and the nuances between their seemingly similar locations proved formative to the development of Chatto's outlook: 'The disastrous results make you think you have a "problem" situation when all you have is a site that different species of plant would enjoy.'[3] Her garden at Elmstead Market in Essex afforded her ample opportunity to get to grips with a range of different conditions – including dry, damp, ultra-dry (gravel) and shady woodland – and to learn which plants best thrive in each of them: 'I am lucky to have room to experiment. Not all gardeners have such a wide canvas as I have here, but most gardens consist of sunny and shady areas requiring different treatments and plants suited to the conditions . . . Success depends on some knowledge of plant provenance and on an understanding of natural plant associations.' Her ideas were consolidated by the knowledge acquired through her husband Andrew's lifelong study of the geographical origins of garden plants and their natural associations in the wild, and through further trips abroad: 'Perhaps the best place to seek knowledge is to go and see the plants growing in their native environment.'[4]

The importance of this was considerable: 'I think for us both the significance of seeing plants growing together in association and in different situations was what really fired us with the idea of making a garden based on ecological ideas.'[5] Her understanding of these not only transformed her garden, it also helped her establish a ground-breaking nursery which both introduced a wide range of unusual plants to the public and also helped to disseminate an ecological sensibility within British gardening. In her displays in the Floral Marquee at the RHS Chelsea Flower Show in the 1970s, she grouped plants according to their natural settings, as opposed to the normal procedure of placing them according to decorative whim. The move was somewhat controversial at the time, although it landed her a string of gold medals and the adoration of the horticultural cognoscenti with a penchant for unusual plants.

Her burgeoning plant catalogue was based upon her experiences of growing plants as naturalistically as possible:

> During the last 50 years, there has been emerging, like a tidal wave, a surge of interest in species plants – plants as they are found growing in the wild . . . Vast numbers of species plants have been, and are still being, introduced . . . but often on receiving a new plant we find we know little or nothing of its needs, and nothing at all about the plants it was growing with. Was it sheltered in bushy scrub, in deep black meadow soil or on a sunbaked bank? This information would help all gardeners and their choice of appropriate plants.[6]

However, aware of this as she was, she nonetheless appreciated the artifice involved in garden-making: 'I try to follow Nature (not to copy her; we cannot do that in a garden), putting together plants which have similar needs in a situation for which they are adapted.'[7] This balance was articulated through distinct areas of planting within her garden, including woodland and damp gardens. The woodland garden proved to be an expansion and refinement of the wild garden idea, featuring a wide-ranging palette of introduced plants naturalized with natives (Plate 6).

Acknowledging the low rainfall conditions of East Anglia, Chatto created a gravel garden on the site of a former visitors' car park with an assortment of plants from drier and Mediterranean areas (Plate 7). The arrangement provided amenable conditions for the plants deployed, although the design in island beds still compiled them in a fairly traditional manner, as opposed to recreating naturalistic spatial formations, and the plant pairings were arranged according to aesthetic criteria. The influence of the gravel garden cannot be underestimated in its effect on the gardening public, in getting people to recognize that the creative opportunities for planting are closely aligned with the ecological parameters in which plants thrive. Yet Chatto was insistent that while there were certainly take-home ideas to be garnered from the garden, it was not possible to simply imitate it: 'My interpretation of the gravel garden is specific to my soil and conditions – it is not any piece of soil filled with an assortment of unrelated plants then covered with stones for effect. And although I grow a wide range of plants which can be grown in similar conditions around the temperate world, I must emphasize the word similar, that is, much higher temperatures or far lower winter temperatures than we have would limit the range.'[8] This attention to the environmental and climatic conditions moved her far beyond the thinking inherent in the garden scene surrounding her, something she was keenly aware of: 'The point I need to stress is that copies of my gravel garden will not necessarily be successful or suitable if the principles underlying my planting design are not understood.'[9]

The rupture of her ideas from traditional horticultural approaches to planting was best evidenced in her relationship to colour. In *Beth Chatto's Green Tapestry* (1989), she fired a broadside against the traditional colourists and their ardently aesthetic approach: 'I would never deliberately plan a white garden for instance. If you do, the danger is that you are going to take white plants requiring very different conditions and try to grow them all in one area. That goes against my theories of gardening.'[10] Expanding further and getting more specific, she railed in firm manner: 'I am not drawn to a garden planned in this way, I find it artificial. It is heresy to say so, but the White Garden at Sissinghurst is not something I can totally admire . . . I am not sure that I liked it, and I certainly did not feel at home in it, even though I was very interested in the individual plants and excited by the great variety of white-flowering ones.'[11] And, in a dismissal of the Jekyllian foundations of British gardening: 'I am never concerned if the colour has gone out of a border. I would not rush in to remedy the lack with begonias and dahlias. My style of gardening does not depend too much on colour and, where it does, it is in a more subtle way.'[12]

Also in *Green Tapestry*, the formality of garden rooms and evergreen structure provoked her to dismissal of another English icon: 'Although I know they are an integral part of classically styled gardens, I am not keen on clipped box hedges forming a barrier between me and the plants they enclose . . . Take a garden like Hidcote. Although I admire the combination of colours, the planting and design, with the compartments of the garden spread like rooms with Persian carpets of plants, I would never choose to make a garden like that.'[13]

For Chatto, colour was simply one aspect of a planting arrangement: 'I cannot emphasize enough that the form and shape of the plant and the texture and colour of its leaves are as important as the colour of its flowers.'[14] Her attention was holistically focused on the entire plant and the balance it creates within its association with other plants, so that 'overall there is a balance of harmony, of shape, form, outline, and texture.'[15]

Chatto's ecological aspirations included understanding the appropriate properties of soil and their relationship to temperature and moisture, as a key to creating a garden that would not be reliant on regular watering. Yet intervention in establishing this balance was still something she considered acceptable, in so far as she was prepared to improve soil every six or seven years, even if this was antithetical to a truly ecological perspective. The conditions of plant communities were key to her approach rather than all of the interlinked processes that produced them. She was aware of the distinction between a garden and a natural area of plant growth: 'When it comes to putting it into practice you cannot be a complete purist. Gardening is the art of combining plants from many different areas of the temperate world, to provide pleasure and interest for a much longer season than, say, that of the flowering meadows of the Swiss Alps. Furthermore, there are many plants, such as stinging nettles or running twitch grass, that may be part of a natural association but are not desirable in the garden.'[16]

Chatto's garden was a living exposition of the ideas contained within her books: *The Dry Garden* (1978), *The Damp Garden* (1982), *Beth Chatto's Gravel Garden* (2000)[17] and *Beth Chatto's Woodland Garden* (2002),[18] all of which presented her ecological understanding of plants within an approachable and recognizable horticultural context. These ideas, along with a keen awareness of environmental vicissitudes, noted from her meticulously regular recording of rainfall, provided prescient insight into the challenges that lay ahead for gardeners: 'Wherever we live, we all have to assess the amount or lack of water, the degree of heat and cold we receive. Whether or not we take global warming seriously, weather patterns are changing.'[19]

PLANTS AS PROCESSES

ECOLOGICAL PLANTING

Chatto's 'right plant, right place' method of planting, informed by plant associations in their natural habitats, suggested the transition from a traditional way of thinking about plants towards a more nuanced and sensitive approach based on a close understanding of their needs. Her influence has been acknowledged by garden designer Dan Pearson, whose work in domestic and public projects exhibits an astute awareness of the needs of plants, while gardeners Keith Wiley in Britain and Peter Korn in Sweden have pushed the 'right plant' logic to the extreme by creating the 'right place' for them through large-scale efforts of land forming and substrate manipulation.

But this method was actually nothing new: it had a history that had begun a century earlier than the Essex plantswoman's achievements in the gardening world. This particular history developed primarily in Western countries with established garden cultures, as people began applying new scientific findings and theories to move beyond standard horticultural practice in novel ways.

Ecologically based planting has provided an alternative paradigm to that of traditional gardening, proposing gardens as processes rather than pictures. The typical notion of a garden as a static assembly of plants intended to remain looking similar year after year requires ongoing effort and considerable resources in order to satisfy the culturally determined aesthetic criteria, and denies the underlying dynamic impulses that power plant life. The wish to maintain an ideal portrait of plants is in direct conflict with the dynamic nature of vegetation, and reflects a perspective that is egological rather than ecological.

A great deal of effort and many inputs are necessary to keep plants in such an idealized state: it manipulates place to meet the needs of plants by relying on an unsustainable cycle of fertilizing and removing any unwanted intruders. From initial site modification by the addition of soils and nutrients through to never-ending weeding, feeding, watering and pruning, this form of gardening is time intensive, costly and resource hungry. But without this continual support system most gardeners wouldn't be able to achieve their intended goal. The ongoing cycle of investment in intervention makes plants vulnerable to disturbance and lacking in natural resilience, and as the overall plantings are less resistant to intrusion from unwanted or unintended species, increased vigilance is required. Add to that the challenges of plants surviving through a changing climate, and the task to ensure that gardens are picture-perfect becomes ever more arduous and further removed from the Edenic dream or, indeed, contemporary ideas of 'connecting with nature'.

As garden ecology highlights, any style of planting can offer a variety of environmental benefits, but because traditional horticulture relies upon a series of ecological simplifications that reduce the diverse differences of plant needs into a few simple categories, it often negates or forecloses the possibility of other positive benefits that could be achieved. Nearly every horticultural website or nursery plant description provides very general information, listing soil type and pH, moisture and light availability as the main plant needs to be considered, along with their aesthetic qualities of colour, form, texture and seasonality. These categories are extremely broad and gather together disparate plants into similar groupings, which neglect to take into account more specific factors that are the result of co-evolution with the environment and are determinants of plant health and survival. Plants have adapted to live in idealized niches which meet all the criteria for their needs. Because of environmental variations these are not always fulfilled, and they instead live in a realized niche which meets enough of them. The plasticity of plants in their ability to accommodate such variations provides them with the capacity to live in a range of different situations. As a consequence, they have the ability to tolerate a wide variety of garden conditions, thereby making the recommended planting conditions seem like they are a perfect fit – yet to ensure this is so entails constant manipulation of the environment. While Chatto's method for plant selection pays greater attention to these habitat details, ecological planting digs deeper into the abiotic and biotic environmental factors that affect a particular species of plant in its natural habitat. This form of study, known as autecology, provides greater detail about a plant's requirements and behavioural characteristics, which is essential knowledge for ensuring a better fit to an intended location on a site.

At the heart of the majority of ornamental planting styles is the tendency to consider plants from an individual perspective, whether they are chosen for functional or for aesthetic purposes based upon their physical characteristics, be they evergreen shrubs for structure or flowering perennials for pointillistic punctuations of colour. This view of plants treats them as discrete entities that can simply be slotted in wherever desired – provided that they also fulfil the generalist site criteria – in a modular approach, the logic of which is reinforced by their objectification as consumer commodities. The problem with viewing plants individually to be placed in random assemblages and not as groups with natural associations is that, again, it is denying the ecological reality of how plants exist.

Ecological approaches to planting look beyond this narrow perspective and deal instead with overall communities, in which plants are continuously engaged in interactions with each other as well as responding to other external influences. Most plants are members of populations of species living together with other populations in communities. The plant community is central to the discipline of plant ecology, and refers to a collection of plants within a specific geographical area which can be defined by its physiognomic properties or its floristic make-up. Plant communities function according to the ways in which each species within them interacts with environmental conditions and the other inhabitants. Communities differ in their species make-up according to various geographic and climatic determinants, with distinct types of assembly and patterning depending on their spatial layout and density and the abundance of species within them. They are often defined by a dominant species which has a guiding influence on its overall composition and function, such as grasslands or woodlands.

Understanding the interactions of plants within the context of communities presents a more nuanced and complex picture of plant life, and applying this knowledge when designing planting schemes ensures that plants are more likely to survive and, when established, to thrive. The study of synecology – the relationships between plants that affect the composition and function of plant communities – has been a key driver of plant ecology, developing as progressive research has revealed ever more insights into the complex interactions between plants and their surroundings. The history of plant ecology reveals a discipline of contested theories around ideas of change and stasis, instability and equilibrium, and competition and collaboration, and has been instrumental in informing methods of planting.

PLANT COMMUNITIES

A plant community is a group of plants that simultaneously coexist in the same place, dynamically interacting with each other in ways that affect the diversity and abundance of the populations of different species within it. They are typically defined by the relative uniformity of their form and the species they contain, as well as the ways in which they develop through time. As a scientific discipline, plant ecology emerged from Germany and Scandinavia in the late nineteenth century, and has focused on discerning the nature of communities, looking at their composition to determine whether they are distinctly determined groups with little variation or random configurations assembled by environmental factors over time.

The foundation of modern ideas about plant ecology began with the German naturalist Alexander von Humboldt. Rejecting the fascination with individual plants which occupied botanists at the time, he believed that beyond their external features, they were socially organized according to underlying interconnections, a concept he called 'Naturgemälde'. Influenced by the ideas of fellow German botanist Karl Ludwig Willdenow, his travels in South American confirmed a number of important suppositions about the relationships between plants. His 1807 *Essai sur la géographie des plantes* (Essay on the Geography of Plants), co-authored with Aime Bonpland, mapped and categorized the geographic distribution of plants, noting the geological and climatic conditions they existed in. He recognized that globally there were recurring biogeographical patterns and similarities in plant physiognomies – the general appearance and growth habits of plants growing together – in what he informally called 'associations'. These were found to systematically occur in different locations at the same latitudes. They also exhibited the same characteristics at high altitudes on a mountain as they did at lower altitudes further towards the poles, suggesting that despite their spatial separation, there was a correlation between factors such as air pressure, temperature and rainfall conditions, and the type of plants that grew in them. He also observed that plant species diversity increases from the poles to the equator.

Danish botanist Johannes Eugenius Bülow Warming developed Humboldt's ideas a step further, moving beyond a merely descriptive approach to investigate in detail the relationship between vegetation and its environment by taking into account the influence of both biotic and abiotic factors that shape them. In his 1895 book *Plantesamfund*, translated into English in 1909 as *Oecology of Plants: An Introduction to the Study of Plant Communities*, he sought to understand why species each have their own specific habitats, why they arrange in communities, and why these assume distinct forms. On the basis of observations in Brazil, Denmark, Norway, Finland and Greenland, he investigated why genetically distinct species in similar conditions but different regions of the world would adapt to challenges such as drought, flooding, salt and temperature in similar ways. His understanding of how these factors affected individual plants led to the idea that these influences also had consequences for the manner in which different plants interacted with each other in the community.

While this early research began to build a framework for understanding the distribution of communities, the work of American botanist Henry Cowles explored the dynamic aspects of their long-term life cycles. In his 1898 dissertation, 'An Ecological Study of the Sand Dune Flora of Northern Indiana', he put forward the idea of plant succession, proposing that over time plant communities move from a simple level of organization to greater complexity. While changes are continuously happening within communities, such as when new plants replace dead ones and when seasonal fluctuations affect them, succession takes place at a larger scale, usually from one hectare to several kilometres, and within longer time frames, from several to several hundred years. Primary succession occurs where there is no vegetation in the area to begin with and the area is then colonized, while secondary succession entails an initial plant community that is transformed into another type of community, such as a grassland into a woodland. But the nature of how succession actually progressed proved to be a contentious matter for some time.

In 1916, American plant ecologist Frederic Clements suggested in *Plant Succession: An Analysis of the Development of Vegetation* that this process was a unidirectional one progressing through a defined series of stages leading to a final stable climax state, determined primarily by the local climate. He considered that the plant community acted as a singular superorganism living through a predetermined life cycle, in which its species membership, overall structure and patterns could be anticipated at each stage of the process. This proceeded according to a set sequence of stages. First, nudation occurred when a bare area was created by some form of disturbance, followed by migration, which heralded the arrival of plants. Ecesis saw their establishment, with coaction following, involving interactions and competition between species. Reaction involved the modified site facilitating the growth of other species, and finally stabilization was achieved as the fixed end state. This process would then be repeated, with any deviation from it attributed to non-ideal conditions. Clements considered the final climax community to be perfectly adapted to its climate and environment. He considered communities to be homogenous and distinct from each other, with abrupt transitions separating them.

Clements's theory went unchallenged until 1926, when ecologist Henry Gleason disputed its deterministic and uniform aspects in *The Individualistic Concept of the Plant Association*. He instead proposed that plant associations were the product of individual plants coalescing in a random manner, because

> every species of plant is a law unto itself, the distribution of which in space depends upon its individual peculiarities of migration and environmental requirements. Its disseminules migrate everywhere, and grow wherever they find favorable conditions. The species disappears from areas where the environment is no longer endurable. The behavior of the plant offers

> in itself no reason at all for the segregation of definite communities. Plant associations, the most conspicuous illustration of the space relation of plants, depend solely on the coincidence of environmental selection and migration over an area of recognizable extent and usually for a time of considerable duration.[1]

Gleason's theory failed to receive a positive reception and left him as something of an outlier in the field, only really gaining critical appraisal a few decades later with the appearance of other theories challenging Clements's hegemony. Plant ecologist Robert Whittaker demonstrated in his 1948 dissertation, 'A Vegetation Analysis of the Great Smoky Mountains', that populations and communities were independently scattered, with the boundary areas between them – known as ecotones – ill-defined and gradually blending into one another to form a continuum. He stated, 'It is time for community ecology to be based on gradients and distributions when practicable and for the theory to be reconsidered from relative and individualistic viewpoints.'[2] He recognized the usefulness of breaking this continuum into sections for analytic purposes, but these were abstract cultural categories, not naturally occurring ones. The gradual changing of species along the gradient was determined by the various changes in geography, climate and topography: 'If distributions are observed in nature apart from preconceived associations and along more than one gradient, it can hardly be denied that no two distributions are alike, and it is only reasonable to assume that every species is distributed independently according to its own physiology, reproduction and survival, and competitive ability.'[3]

Plant ecologist Frank Egler challenged Clements's view in his 1954 article, 'Vegetation science concepts 1: Initial floristic composition, a factor in old-field vegetation development'. His theory, based upon disused fields in the United States, postulated that all the species that develop on a site are present at the time it is abandoned, rather than appearing sequentially as part of a preordained process. This site-specific development proceeds at different rates according to the attendant species:

> In this case, up to the year of abandonment, the land is receiving many species, as seeds and living roots. Ploughing and tilling serves to break up and scatter these propagules, and to plant them below the surface . . . After abandonment, development unfolds from this initial flora, without increments by further invasion . . . The annual weeds with their rank growth are first in evidence, with other plants occurring as seeds or seedlings. The annual weeds drop out of the race, and the grasses assume predominance, with woody plants present but small or dormant. As each successive group drops out, a new group of species, there from the start, assumes predominance. Eventually only the trees are left.[4]

Botanist John Curtis's studies of the vegetation of Wisconsin also supported Gleason, although with certain caveats. In *The Vegetation of Wisconsin: An Ordination of Plant Communities*, published in 1959, he cautioned against fully adopting the randomness Gleason proposed, and suggested instead that there is a certain pattern to the vegetation, with more or less similar groups of species recurring from place to place. The main reason for this is the great potentiality for dominance possessed by a relatively small group of species. These are plants that are well adapted to the over-all climate and soil groups of the province and which have the ability to exert a controlling influence on the communities where they occur because of their size or their high population densities . . . The interactions of this tiny group produce a series of microenvironments which differ according to the biological characteristics of the dominants. Most of the remaining species of the flora must grow in these modified conditions and they tend to be sorted out in groups aligned with the particular dominant concerned. The groups are not discreet and separate from one another but gradually shift in composition because the dominants themselves rarely grow in pure stands but rather in mixtures of varying proportions.[5]

Moving the focus further towards the influence of characteristics of particular species, ecologists Joseph Connell and Ralph Slatyer proposed in 1977 that succession was dependent on the particular ruderal pioneer species that first colonized a disturbed environment and the ways in which they went on to transform it, thereby either facilitating or inhibiting later successional species. Alternatively, they might not change the environment significantly enough to promote or prevent other species, resulting in a community consisting of tolerant species that can coexist with other species in a densely populated area. A few years later, Slayter with Ian Noble also suggested that vital attributes of plants predicted their performance in succession premised on the means by which they arrived after disturbance, their ability to enter an existing community successfully and their growth rates.

Positioned between the holistic and individualistic poles, plant ecology today recognizes that numerous possibilities exist rather than one overarching explanation, reflecting the inherent complexity of plant traits and variable external forces operating on plant communities. Ideas of stability and equilibrium have been accepted as being relative, and a wider appreciation has been accorded to heterogeneity. Recent research suggests that communities are driven by process rather than end point, and tend to exist in a recognizable form for between 230 and 460 years, until encountering some form of significant disturbance, and then reconfigure themselves over the next 140 to 290 years, with only 64 per cent of them tending to revert to their previous state. This suggests that not only do plant communities perform various successional processes in which distinguishable forms are identifiable, but they can also change completely into another type of community in the same location, starting a new sequence of changes.

THE COMPETITIVE EDGE AND BEYOND

Another important area of research has been on the characteristics and strategies of plants, which determine the relationships within communities and thereby their composition. A key notion has been the ability of certain species to succeed because of their competitive tendencies, which can reduce fitness in their neighbours due to the shared use of a limited resource.

Rather than thinking of plant communities as a unified successional progression to a predetermined end state, Scottish botanist Alexander Watt instead proposed that they were actually a mosaic of patches dynamically related to one another, balancing predictable structure with volatile random disruptions in many disparate ongoing cycles of recomposition. Rather than assuming that soil and climatic conditions were uniform across the community, his nuanced view assumed that in the different phases of patches, the species within them modify these conditions, and in turn are then modified by them in a feedback loop. This ebb and flow was therefore determined by a combination of soil and microclimatic influences, the competitive traits of the species in the community, and their different life cycle stages. In an article titled 'Pattern and process in the plant community', published in the *Journal of Ecology* in 1947, Watt explained, 'Each patch in this time-space mosaic is dependent on its neighbours and develops under conditions partly imposed by them. The samples of a phase will in general develop under similar conditions but not necessarily the same, for the juxtaposition of phases will vary. This may be expected to affect the rate of development of the patches of a phase and their duration.'[1]

In his work in the late 1980s, American ecologist David Tilman suggested, 'A variety of such models of competition for a limiting nutrient predict that the species that can reduce the concentration of the limiting resource to the lowest level should competitively displace all other species . . . and that the differences among these species are based on allocation.'[2] Therefore the best competitors in a particular environmental situation will be dominant because these species will have evolved by trading off certain physiological or morphological traits in order to enhance other more favourable ones that give them a competitive advantage.

However, ecologist J. Philip Grime, at the University of Sheffield, found that simply limiting interaction to competition alone neglected to consider other important plant traits. His CSR strategy, which he began developing in the 1970s, has become the theory that has had the most impact on current practitioners of ecological planting. It is based on the premise that there are two primary factors that affect plants: stress and disturbance. Different plant species have evolved particular adaptive strategies, which they use to deal with these challenges. In their responses to them, they face a trade-off in the extent to which they invest resources in growth and biomass, capacity for further resource acquisition, or regeneration. The theory can be used as an analytic tool for viewing existing vegetation, but, more importantly, since it can facilitate the ability to predict a species' relative success in different environments and with other species in communities, it can also be employed as a tool for planting design.

The first factor, stress, involves conditions that restrict growth such as the availability of light, water, nutrients and space. The second, disturbance, refers to anything that involves the partial or total destruction of the plant. These determine the structure and composition of communities in different conditions: 'In the spectrum of plant habitats provided by the world's surface the intensities of both stress and disturbance vary enormously. However, when the four permutations of high and low stress with high and low disturbance are examined, it is apparent that only three of these are viable as plant habitats. This is because, in highly disturbed habitats, the effect of severe and continuous stress is to prevent a sufficiently rapid recovery or re-establishment of the vegetation.'[3] Consequently, there are three main categories of plant strategies, which are defined as competitors, stress tolerators and ruderals (hence CSR).

Competitors are plants that have the ability to exploit resources more effectively or efficiently than other neighbouring plants in a community, and therefore have a tendency to dominate. They thrive in nutrient-rich conditions and in low-stress and low-disturbance situations; indicators of competitor species are rapid growth rates and high productivity. Stress tolerators are able to withstand demanding environmental conditions in which resources are limited, such as extremes of temperature and the over-abundance or lack of moisture and light. These species often feature distinct morphologies to enable them to endure the conditions that they exist in, such as cacti and succulents, and are found in high-stress and low-disturbance situations. Ruderals are often short-lived species that reproduce rapidly. With opportunistic tendencies that allow them to pioneer recently disturbed environments, they are early colonizers of spaces and tend to be found in low-stress and high-disturbance situations. They are indicative of the first stage in successional processes, are often annuals, and exhibit quick and prolific reproductive cycles.

Beyond these three poles, there exists a gradient of positions in which the characteristics of various species exhibit a combination of tendencies, referred to as secondary strategies. The four main types are competitive ruderals, stress-tolerant ruderals, stress-tolerant competitors and CSR strategists adapted to habitats where competition is restricted by moderate degrees of both stress and disturbance. Also, an uncoupling of juvenile and mature strategies reveals that plants can change them at different stages of their life cycle.

Grime's book *Plant Strategies and Vegetation Processes*, published in 1979, outlined the theory in detail and ascribed CSR categories to many British species; a more elaborate and expansive categorization was provided in *Comparative Plant Ecology* (1988), and the mathematical calculations underlying the system were described in detail in the research paper 'Allocating C-S-R plant functional types: A soft approach to a hard problem' (1999).

In ecologically designed plant communities, it is important to understand the stress and disturbance factors present in the site in order to know how competitor, stress-tolerator and ruderal species will relate to it, and thus to create resilient balanced plantings. Situations of moderate degrees of stress and disturbance are considered to be the sweet spot most conducive to creating communities that are diverse and relatively stable, as they moderate each of the CSR tendencies of the plants within it. Also, taking into account strategy variations across plant life cycles begins to build a dynamic picture of plant interactions in communities over time.

PLATE 1 Humphry Repton's 'improvements' at Sheringham Park, Norfolk, using trees as framing devices for vistas and also to conceal parts of the landscape.

PLATE 2 Illustration by Alfred Parsons from William Robinson's *The Wild Garden* (1870), showing colonies of poet's narcissi and bergenia, highlighting Robinson's strategy of naturalizing introduced plants in the wild garden.

PLATE 3 The classic Jekyllian border, as pictured in Gertrude Jekyll's *Colour in the Flower Garden* (1908), consisted of long, informally planted flower beds, usually framing a lawn path and filled with a mix of bulbs, annuals, perennials and shrubs arranged to perform at certain times of the year, flowering for a few months each in succession.

PLATE 4 The colourists' penchant for monochromatic planting was the design rationale behind the Red Borders at Hidcote.

PLATE 5 The Long Border at Christopher Lloyd's garden at Great Dixter exhibits the horticultural highpoint in the art of exuberant and extravagant planting.

PLATE 6 The woodland garden at Beth Chatto's garden exemplifies her 'right plant, right place' approach based on natural plant associations.

PLATE 7 The gravel garden created by Beth Chatto on a former car park was planted with a variety of drought tolerant plants to survive in the dry conditions of East Anglia. In considering the importance of environmental conditions, it broke free from the taste-based planting design dictates of the 1990s.

PLATE 8 Taking a cue from natural habitats, ecologically informed planting in layers creates a dense tapestry with plants filling every available niche.

PLATE 9 The Curtis prairie restoration project at the University of Wisconsin–Madison Arboretum was a pioneering project of modern restoration ecology in the United States, created in 1936 under the guiding eyes of ecologist Aldo Leopold and plant ecologist John T. Curtis.

PLATE 10 Brooklyn Botanic Garden Native Flora Extension by Darrel Morrison draws upon native plant communities to recreate a pine barren ecosystem.

PLATE 11 Darrel Morrison's prairie-influenced planting at the Lady Bird Johnson Wildflower Center in Texas.

PLATE 12 Large-scale schemes like the meadow at Wolf Trap National Park in Virginia, designed by Phyto Studio, allow visitors to immerse themselves within the planting to see how plants actually grow together in communities.

PLATE 13 East Coast native plants in Baltimore's Oriole Garden, designed by Claudia West of Phyto Studio, create a biodiverse community providing colour to engage the public in an urban setting next to the local baseball stadium.

PLATE 14 Plant communities illustrated in *De Bonte Wei* (The Colourful Meadow) by Dutch author, teacher and biologist Jacobus P. Thijsse, who introduced the idea of semi-natural parks with a variety of habitats for educational purposes.

PLATE 15 The heempark Jac. P. Thijssepark in Amstelveen is the most well known of the Dutch 'habitat parks'. Instigated by landscape architect Christiaan Broerse and botanist Jacobus (Koos) Landwehr in 1939, on 2 hectares of former agricultural land, it features a series of meadow, heath, woodland, dune and bog, habitats.

PLATE 16 A naturalistic weave of plants on the former nursery site at the Oudolfs' garden Hummelo, featuring an informal dense knit of grasses, native and introduced perennials. It includes seeded cool season European meadow species intermingling with planted taller North American species.

PLATE 17 A classic Piet Oudolf planting of blocks of perennials at RHS Wisley, exemplary of what became internationally known as the New Perennial style.

PLATE 18 The naturalistic woodland at Priona, the garden created by plantsman Henk Gerritsen and his partner Anton Schlepers in Schuinesloot.

PLATE 19 Introduced and native perennial plants mingle among topiary shaped yew (taxus) and box (buxus) at Priona, reflecting Gerristsen's botanical acumen combined with a sense of humour.

PLATE 20 German garden architect Willy Lange's garden in Berlin-Wannsee in 1920, showcasing his idea of fusing ecology and aesthetics to create the Nature Garden.

PLATE 21 Dry prairie planting at Hermannshof, Weinheim. The garden is based on Richard Hansen's Lebensbereiche system, and researches the relationships between plants under various cultivated conditions to develop aesthetically pleasing planting combinations and appropriate management strategies.

PLATE 22 Landscape architect Bettina Jaugstetter has employed tested perennial mixes in public and commercial projects in Germany.

PLATE 23 Mixed perennial planting by Bettina Jaugstetter at ABB company in Ladenburg.

PLATE 24 Teufelsberg in Grunewald, on the outskirts of Berlin, gradually evolved as a 120 m high hill, formed by rubble exported from the inner parts of the city between 1950 and 1972. The site was both planted and allowed to develop spontaneously.

PLATE 25 Natur-Park Schöneberger Südgelände in Berlin, on the site of the former Tempelhof Station switching yards, is managed to ensure successional diversity and the maximum range of plant communities.

PLATE 26 The creation of Park am Gleisdreieck in Berlin incorporated elements of existing spontaneous vegetation into a more obviously designed landscape, providing a practical structure for users and a contrasting framework for wild plants.

PLATE 27 The native meadows and stitch plantings designed by Nigel Dunnett and James Hitchmough at the Queen Elizabeth Olympic Park in London create a series of novel ecosystems exhibiting a unique sense of naturalism.

PRINCIPLES AND PRACTICES

Understanding and applying the principles of plant ecology, at both micro and macro levels, sets ecological planting apart from horticulture. This approach aims to create varied and complex environments for the benefit of the maximum amount of biodiversity in order to achieve healthy functioning ecosystems. Science provides it with a solid foundation for creating dynamic plant communities that are relatively stable yet are able to change through time, have increased resilience and require minimal intervention.

Observing the phenological traits of different species through the seasons and different parts of their life cycles, as well as understanding how they interact as community players, are essential. Thinking about plants according to these short- and long-term behaviours is more important than simply focusing on aesthetic details of colour and texture, although ecological planting takes these into account as part of the wider strategy, and still delivers a high visual impact. Getting the right fit between plants and the environment and finding their appropriate places in a community are learnt from research and observation of their patterns and processes in their natural habitats. This provides informative and instructional cues to how they may live within a designed scheme without being prescriptive or limiting.

Plants have different life cycles, whether annuals or perennials, deciduous or evergreen, that affect the structure of the community throughout the seasons, and how it looks and functions. The inclusion in designed planting of groups with differing lifespans, and at different stages of their lives, mirrors the reality of naturally occurring communities, and the ebb and flow of species within them. These relative fluctuations affect the abundance of populations but don't generally have consequences for the overall community structure, which is determined more by CSR ratios and succession. Designing with these processes in mind and providing for moderate growth scenarios can help build adaptive and resilient communities. As well as competitive tendencies and growth patterns, plant reproductive strategies are another consideration to be factored into the design process, whether clonal or based on pollination and seed distribution through complex relationships with other organisms.

On a more detailed level, by paying close attention to localized microclimates and slight differences across a site due to soil conditions, light and moisture limits, as well as the ways in which plants alter their own environments over time, ecological planting can take into consideration the needs of species that allow them to successfully occupy their niches. Filling as many niches as possible within a community maximizes diversity, and helps to prevent the intrusion of unintended species with overly competitive tendencies, which are likely to be disruptive to ecosystem functions.

A key spatial component of ecological design is creating layers of plants, which provides a three-dimensional way of allowing a diverse range of plants to coexist in a habitat. The idea is drawn from the way plant communities function in the landscape, where plants of different biomass shapes and sizes above ground, and root structure below, fit together and function in compatible ways. A woodland is a prime example of ecological layering: a vertical structure consisting of the upper canopy of trees, mid-height shrubs, smaller perennials and ferns, diminishing down to ground-cover plants, mosses and lichens. Deciduous tree canopies provide shade necessary for the plant species that have evolved to live in such communities. Their phenological clocks utilize the early spring light to grow and perform their reproductive cycles before the sun's energy is shut out by the emergence of the tree leaves. These layers are obvious not only in communities like woodlands, but also at smaller scales and in less apparent ways, such as grasslands and meadows, where early species such as primroses are soon covered over by grasses and perennials with the arrival of more light and warmer weather, yet are adapted to survive and reappear the following year. The use of layers in planting design allows the build-up of a complex network of plants whose characteristics are employed to create the best conditions for the plants themselves, and also to aesthetically transition through seasonal and longer-term cycles. Ground-cover, medium and canopy layers are often used as a basic framework, but drilling down further into specific plant niche requirements affords opportunities for a more complex and nuanced level of vertical composition (Plate 8).

While science is a guiding star, plant ecology has revealed that despite the distinct processes evident in plant communities, their behaviour is not simply mechanistic or wholly predictable, so embracing randomness is a crucial requirement of these new approaches to planting. Having the flexibility to respond to unforeseen circumstances and chance occurrences means that living with ecological planting schemes is an ongoing process and not set in stone at the drawing board. Gardeners are well aware that their best laid plans can so easily be ruptured by an unexpected event, be it weather conditions or the appearance of disruptive animals, insects or pathogens, which undermine the perfection they are aiming for. Within an ecological perspective these are to be accommodated as opportunities to learn from rather than moments of despair. Observation of how plant communities develop is part of the feedback loop in a symbiotic process of management or stewardship, rather than traditional methods of maintenance that attempt to restrain change through overt measures of control.

The visual appearance of ecological community-based planting can be as bright, colourful and arresting as traditional horticultural styles, but the forms it takes, often based upon matrix-style layouts, can carry with them loaded associations of natural environments and landscape ideals. These can evoke a wide variety of cultural connotations and personal interpretations. A naturalistic looking planting may not prove to be suitable for a densely built and highly populated urban area in some situations, whereas in others it may be welcomed as providing a striking and rejuvenating contrast. Landscape architect Joan Nassauer noted in 1995: 'Applied landscape ecology is essentially a design problem. It is not a straightforward problem of attending to scientific knowledge of ecosystem relationships or an artistic problem of expressing ecological function, but a public landscape problem of addressing cultural expectations that only tangentially relate to ecological function or high art. It requires the translation of ecological patterns into cultural language.'[1]

Naturalistic looking planting has generally been designed to ensure that there is some indication it is intentional and not simply spontaneous and therefore unconsidered. The design element can be subtle, yet show that the planting is deliberate and, as a consequence, less threatening or undesirable. Framing plants within a readable context provides them with 'cues to care'.[2] A sense that a landscape is cared for generates a sense of acceptance and respect, so ensuring that these cues are legible in a planting design is important.

Increasingly diverse demographics mean that considered attention to cultural differences needs to be given in order to be inclusive and maximize the benefits of planting socially as well as environmentally. Much research on the socio-psychological appearances of planted landscapes has been carried out at the University of Sheffield, which reveals wide-ranging interpretations of naturalistic looking planting that depend on demographic factors including age, gender, ethnicity and lived experience. One field study paper suggests, 'Further research needs to focus on understanding the role of gender and the values and behaviour of people who do not demonstrate nature-connectedness and do not visit or use these spaces, particularly in the global south.'[3] Appreciating these varied viewpoints as another aspect of site conditions can then help inform decisions regarding plant selection and the form that the plant community may take. People are very much part of the biodiversity mix, and the skill is in achieving this satisfactorily while ensuring it also functions ecologically.

As a consequence, ecological planting is not a one size fits all approach, as the variable factors that act on plant communities mean that it needs to be responsive to a wide range of site conditions as well as intended social outcomes. This is borne out by the wide range of rationales and techniques employed by contemporary practitioners. While basic ecological science is at the heart of what they all do, the technical, aesthetic and cultural differences between them are wide.

One of the first methods developed was a biogeographical approach based upon the fact that plants are well adapted to their locations due to long-term evolution. This takes advantage of plants that have co-evolved together to be compatible, both with the environmental conditions as well as with each other's growth patterns, forms and characteristics. Using locally specific native plants lends them a recognizable spatial and visual coherence through their familiar appearances and floristic composition. This restoration-orientated method has a value where there are key historical or cultural outcomes to be delivered, along with the functional ones the community itself delivers.

Another increasingly popular approach is more flexible, and entails creatively making new associations of plants regardless of geographic origin by matching their needs carefully with the environmental conditions of the site. As plant ecology has revealed, plant origins reflect adaptive traits to local environments but exhibit similarities around the world, so plants from geographically separate regions can be assembled imaginatively into mixed novel communities that account for each species' niche requirements. The aim is to produce a general naturalistic impression rather than a direct facsimile while ensuring that it functions ecologically, so ensuring proximity to the original habitat conditions is key, as it means that plants will be more liable to support themselves with minimal management interventions.

Unfettered in plant choice by restrictions on native or introduced species, this approach can be more practical for building resilient and diverse communities in areas where there is no historically identified vegetation, such as in cities. A well-chosen planting palette can cope with the unique challenges posed by built environments, as well as having a better adaptive potential for future climatic change. It can also provide the opportunity for extended seasons of interest, which can play a part in creating more interesting and engaging places in the public realm, providing social and health benefits.

The different approaches of contemporary practitioners also reflect cultural influences and historic trajectories, which frame the rationale and implementation as well as public appreciation of planting projects. The native plant movement has been a prime mover in the United States, imbuing new approaches with a preservationist outlook. But the scale of the country actually makes the idea of national flora meaningless, given all the different biomes it encompasses, and leads to a more specific regional appreciation of plants and places. With few remnants of 'wild' landscapes left in Europe, the approach is more open and flexible. In the manufactured landscapes of the Netherlands it is decidedly pragmatic, in Britain a scientific empiricism leads the way, in France a philosophical perspective, while in Germany a technocratic rationalism is in evidence. These approaches, while seeming to evoke national stereotypes, are in fact all driven by similar ecological principles, yet they address the fact that complexity requires relative responses, both biologically and socially, in ways to ensure that planting can assist in facing the effects of the climate and biodiversity crises within different cultural contexts. The range of different ideas and methods reinforces the notion at the heart of ecology: diversity is healthy.

ECOLOGICAL DEVELOPMENTS IN THE UNITED STATES

TAKING ROOT: THE BEGINNINGS OF NATIVE PLANT DESIGN AND PRAIRIE CONSERVATION

Parallel to the Robinsonian revolution in Britain in the late nineteenth century (see 'Plants as Pictures'), similar developments began to take shape in the United States. After initial training in the nursery industry and employment as a planting supervisor with pioneering landscape architect Frederick Law Olmsted, Warren H. Manning put his considerable plant knowledge and horticultural experience into practice doing landscape design with more naturalistic tendencies. An appreciation of native plant groupings led to his own take on the wild garden, which involved a considered conservationist approach that aimed to minimize intervention in the creation of gardens. As he outlined in an essay on 'The nature garden', 'I would have you give your thoughts to a new type of gardening wherein the Landscaper recognizes, first, the beauty of existing conditions and develops this beauty to the minutest detail by the elimination of material that is out of place in a development scheme by selective thinning, grubbing, and trimming, instead of by destroying all natural ground cover vegetation or modifying the contour, character, and water context of existing soil.'[1]

While Manning recognized that garden owners were not quite ready to abandon their precious lawns for his radical light-touch approach, he nonetheless encouraged them to get to know local flora in order to observe its intricate details and appreciate its true beauty. He cautioned:

> Never destroy a native tangle of plants without watching for a full year the habits, foliage, flowers, and fruit of every plant growing therein, for if this is done it will be found that many are so attractive at one season or another that they will be retained and developed in beauty by the gradual removal of the less desirable kinds, for which other attractive plants may be substituted that will not interfere with the growth of the permanent plants. A more varied, effective, and interesting group can usually be obtained in this way with less trouble and expense than when the original growth is destroyed, the grounds laboriously prepared, and then planted with expensive young nursery plants.[2]

Manning's design decisions on which plants to retain and which to exclude were taken from an aesthetic standpoint, but clearly marked a step back from traditional approaches. At the Gwinn garden outside Cleveland, Ohio, he showcased his approach to creating the wild garden, making an inventory of the site's existing plants, and then selectively editing and adding, which resulted in a small woodland and a 8.5-hectare/21-acre area featuring masses of rhododendrons, wild flowers and ferns.

A uniquely Midwestern style began to take shape throughout the late nineteenth and early twentieth centuries in the work of landscape architects Ossian Cole Simonds and Danish émigré Jens Jensen. This pioneering approach used a native-based planting style inspired by the natural vegetation of the prairies and savannahs. Concerned with the rapid disappearance of the native flora due to the encroachment of large-scale agriculture, what became known as the 'prairie style' was inspired by natural groupings of native Midwestern plant communities which sat in harmony with the surrounding landscape.

In his 1936 book *Siftings*, Jensen suggested: 'There are trees that belong to low grounds and those that have adapted themselves to highlands. They always thrive best amid the conditions they have chosen for themselves through many years of selection and elimination. They tell us that they love to grow here, and only here will they speak in their fullest measure.'[3] He was convinced that 'every plant has fitness and must be placed in its proper surroundings so as to bring out its full beauty. Therein lies the art of landscaping.'[4] Proper surroundings were geographical for Jensen, and planting should be related to the conditions of the regional area they originated from. The fact that gardens and landscapes were filled with imported species was a matter of consternation for him: 'It is often remarked, "native plants are coarse." How humiliating to hear an American speak so of plants with which the Great Master has decorated his land! To me no plant is more refined than that which belongs. There is no comparison between native plants and those imported from foreign shores which are, and shall always remain so, novelties.'[5]

Jensen was instrumental in establishing the Friends of Our Native Landscape in 1918. Primarily focused on Illinois, the group commissioned an investigation into preserving 'what is left of the scenic and historic lands in this state'.[6] The resulting 1922 publication, *Proposed Park Areas in the State of Illinois: A Report with Recommendations*, identified various 'large areas to preserve the native flora and fauna in all its wild and mysterious beauty. Overcrowded parks or preserves mean the destruction of all such.'[7] The areas included various plant communities, habitats and landscapes such as prairie, forest and river valleys.

He spurned detailed planting plans and instead opted for large-scale groupings of plants massed in intricate patterns, creating a rhythm through repetitions of massed forms and textures, and gradually developed an approach based on plant communities with distinct layers to the arrangements. In an early convergence between planting design and plant science, Jensen's ecologically inspired approach attempted to select plants in accordance with Cowles's idea of succession. The two men were friends and instrumental in founding the Prairie Club in 1908, which encouraged urbanite Chicagoans to engage with natural landscapes; they were also active in the campaign to preserve the Indiana Dunes, which were the basis of Cowles's work. Jensen's mix of design and science attempted to strike a balance between the spatial layout of planting on a site and its changes over time due to dominant species. His most successful endeavour in applying these ideas was at Lincoln Memorial Garden in Springfield, Illinois, a 24-hectare/59-acre park begun in 1933 and planted over many years, which featured a series of meadows separated by lines, each focusing on particular plant communities.

Regional landscapes and planting carried more than just an ecological significance for Jensen; they also held something of a moral imperative. In 1935 he transformed his own garden into The Clearing, a folk school where city people could renew their contact with the 'soil' and, as the name suggests, clear their minds of the detritus of urban life through an engagement with the natural environment. In *Siftings*, he attributed certain characteristics among populations of European countries and American regions to the influence of their landscapes.

Jensen's regional enthusiasm took on somewhat darker tones in the late 1930s when he stated in the German publication *Die Gartenkunst*, 'The gardens I create, like any landscape design, should be at one with their natural surroundings and the racial characteristics of their inhabitants. They are meant to express the spirit of America and therefore should be free from foreign influences.'[8] Standing on shaky ground, Jensen elided the provenance of plants with people, and went on to espouse the racist view that Latin and Asian immigrants spoiled the character of society.

The 'prairie style' had been popularized by horticulture writer and academic Wilhelm Miller's circular 'The prairie spirit in landscape gardening' (1915), which drew upon the work of Jensen and Simonds. He summarized, 'The prairie style of landscape gardening, however, is a genuine style in the opinion of several critics, for it is based on a geographic, climatic, and scenic unit, and it employs three accepted principles of design – conservation of native scenery, restoration of local vegetation, and repetition of a dominant line. However, it is not a system of rules and there never can be anything of the sort in any fine art, tho people crave it forever. In good design there are only principles.'[9]

Aesthetically, Miller referred to the physical openness of the prairie, and in particular its linear flatness, as a general guiding principle to planting, but many of the sites that he discussed, such as woodlands, riverside, roadsides and farms, included quite a diversity of vegetation. Decrying the loss of the local character of the landscape and the native plants that had inhabited it, Miller's aim was less about recreating authentic prairies in their original form than encouraging landowners of urban, suburban and rural plots in the Midwest to enrich their environments with native plants, even exhorting farmers to let brush remain at the sides of their fields. In parks in city areas, his prime consideration was to incorporate prairie plants, even if they were only in a small area.

Miller's outlook was also influenced by Cowles's work, but interpreted in a fairly simplistic way, focusing on native plant communities' evolutionary adaptations to local conditions. His proselytizing Midwest regionalism manifested itself as an aesthetic form of restoration based on two scenic spatial strategies inspired by prairies: 'The broad view is the one that suggests infinity and power, and is the more inspiring for occasional visits; the long view is more human and intimate, and often more satisfactory to live with.'[10] Native trees and shrubs were employed to screen and frame views, and to accentuate a sense of expansiveness, in order to emulate local natural landscape features, while repetition of specific plants ensured continuity.

Advancing a polite attack on gardeners for their well-meaning yet misguided enthusiasm for plants fuelled by fashionable tastes, he critiqued bedding plants, annuals, bright flowers and coloured leaves, an outlook he shared with Robinson, whom he had visited at Gravetye in 1908. A concerted dismissal of the colourist approach to planting was central to the thrust of his argument, alongside a celebration of the seasonal delight of plants rather than the artifice of extended blooming through the year: 'On the contrary, we believe all their motives can generally be reduced to four innocent desires that may be grounded in instinct. For everybody loves flowers and colour; everyone likes to have shade and beauty as quickly as possible; everybody likes a little variety or spice in life; and everyone has at least a rudimentary respect for neatness and order. Is it not possible that most of the alleged vulgarity is simply an excess of these virtues?'[11] Individualistic approaches to planting front gardens based upon these misguided notions were his primary concern. The public realm should reflect natural plants to harmonize with the wider landscape and not be cluttered by visually disrupting imported plants. If gardeners were to continue their fascination with them, Miller suggested that they do so in the back garden, out of sight.

Despite its ecological intentions, the prairie perspective was, however, driven by a romanticized vision of the landscape framed within a colonialist mindset. The revered prairies Miller referred to were those seen by white settlers when they arrived in the Midwest. But for all their naturalistic appearance, these were not landscapes free from human intervention. Instead, they were the homelands of indigenous peoples who had established co-dependent relationships with these environments over long periods of time. The flora within them was not simply a product of plants revelling in wild abandon, but a product of cycles of hunting and bison grazing that entailed periods of disturbance to which the plants had adapted. If there were a failing of the prairie style ecologically, it was the lack of understanding of these factors when translating them in a sustainable way into suburban garden settings.

A parallel perspective to creating natural landscapes was developed by landscape architect Frank A. Waugh. Having spent some time in Germany studying under horticulturalist and garden architect Willy Lange, in 1910 he returned to the United States, enthused with ideas for applying insights from his mentor's biological-physiognomical method.

His outlook was focused on the specificity of landscapes and designing appropriately in accordance with them. He was cognizant of the nuances that occur across such a large landmass as the United States, conceding that landscapes vary from region to region and locality to locality, and that these need to be taken into consideration when using plants. As a consequence, native plants figure largely, yet Waugh, with a comprehensive horticultural background, was pragmatic enough to include what he considered to be worthy introductions suitably naturalized.

More pertinent was the art of arrangement. He was convinced that placing plants together in gardens should be based on the ecological principles of their associations in natural environments, and extolled the virtues of the natural style as a fundamental garden form as opposed to formalist design. Contrasting the two, he noted that the former was defined by a lack of symmetry and unencumbered by enclosure and boundaries. While aware of the contrived nature of the garden, he noted, 'It is a structural form characterized by certain resemblances to the natural landscape. These points of resemblance are sometimes quite arbitrarily chosen by the garden designer, and sometimes quite artificially developed; but it is always the logical aim of the artist to discover and to follow the principles of composition followed by nature.'[12]

A note of Transcendentalism from the nature-reverent philosophy of nineteenth-century essayists Ralph Waldo Emerson and Henry David Thoreau was detectable in the other defining feature of the natural style. In *The Natural Style of Landscape Gardening* (1917), he asked, 'What then is the informing spirit of the natural style? Is it not the spirit of the natural landscape? We speak of the spirit of the woods, or the spirit of the mountains; and, quite as precisely as common language can ever convey spiritual ideas, we know what we mean. We do actually have a perfectly clear idea in mind when we speak of these things.'[13]

The work of design was to comprehend this spirit and translate it by simplifying and accentuating the characteristic natural forms of the landscape. Waugh considered it 'axiomatic that landscape gardening draws its materials and ideals from the landscape . . . the native landscape serves with special efficiency in supplying materials and models for that form of landscape gardening which we call the natural style . . . the materials, the form and the spirit of the open landscape should be sympathetically understood by every landscape architect, no matter how narrowly his practice may be restricted to the most formal work.'[14] Furthermore, landscape architects were tasked not only to interpret the landscape but also to conserve and protect it, and make it accessible to human beings. They were ordained to translate the beauties of the landscape to the general public, in much the same way as a skilled musician would interpret a score by a classical master, something that could only be achieved by an intimate familiarity with it.

The 1929 book *American Plants for American Gardens*, by plant ecologist Dr Edith Roberts and landscape architect Elsa Rehmann, is perhaps the key work of the pioneers of ecological planting in the United States, which laid the groundwork for contemporary practitioners. In it the authors advocated the use of native plants arranged in communities according to recognizable types, such as open fields, juniper hillsides, oak woods, pine forests and stream sides. Countering the colonial history of introduced plants that continued to curry favour with gardeners at the time of writing, they waved the flag for home-grown species. More astute than the contemporary advocates of naturalistic planting, their approach dug deep into the latest available science:

> In plant ecology, observations are made as to what plants grow together and how they compose the groups known as plant associations. Observations are also made of what the plants in each association have in common as to soil, light, moisture and temperature, all of which are the factors which make up what is called the plant's environment. These observations draw attention to the fact that every slight variation in any one of these factors changes the members of each association and brings about the infinite variety that exists in the natural grouping of plants.[15]

Recognizing the dynamic nature of these plant communities and their states of succession, they proposed that gardeners familiarize themselves with plant communities and consider how they could incorporate lessons from them into their own spaces. The authors examined the different type of communities in forensic detail, running through the different species associated with each, relating shrubs and perennials to the dominant trees or defining plants of each community. The descriptions of plant forms and characteristics, suggestions and plant lists were intended as a guide for translating the communities into domestic settings. The proposed method was that if a garden owner already had a particular tree, then they could then add the other plants associated with it into the garden, or alternatively edit out the existing species that didn't fit with the community. This closely observed study of native plants was intended to bring out 'the fundamental principles upon which the indigenous vegetation is established, and the contribution that an understanding of these facts can make to the retention or re-creation of the natural landscape. It draws attention, moreover, to the significance of retaining the original contours, and the adaptation of a house and garden to the lay of the land and to the spirit of the natural landscape.'[16]

BLOOMING: DESIGNING WITH NATURE

Although not particularly plant-orientated, Scottish-born landscape architect Ian McHarg developed a thorough understanding of ecology and ways in which it could be applied in large-scale landscape projects. His polymathic outlook was encapsulated in his landmark publication *Design with Nature* (1969), in which he charted new ways of thinking about landscape design and planning at local and regional scales in relation to climate, geology, hydrology, limnology, soil, vegetation and wildlife.

His awareness of the anthropocentric destruction wrought upon the world manifested in a mixture of frustration, exasperation and disgust with humans and their refusal to grasp the interconnectedness of life processes. His recognition of the growing environmental crisis was prescient: 'Man is natural, as is the phenomenal world he inhabits, yet with greater power, mobility, and fewer genetic restraints, his impact upon this world exceeds that of any creature. The transformations he creates are often deleterious to other biological systems, but in this he is no different from many other creatures. However, these transformations are often needlessly destructive to other organisms and systems, and even more important, by conscious choice and inadvertence, also deleterious to man.'[17]

His understanding of these processes was based on a heady mix of biology, chemistry, anthropology, and garden and landscape history, framed within a critical view of the industrial and economic systems of capitalism that exploit and devalue them in pursuit of profits. His grasp of ecology was timely in the wake of Rachel Carson's exposé of the toxic pollution in *Silent Spring* (1962) and the growing environmental consciousness in the counterculture of the late 1960s.

While McHarg's work focused on large-scale environments, his vision was informed by ideas of nature within garden history. His analytic reflections on the Western garden tradition revealed the predominance of aesthetics and infatuation with all things floral to be the determining aspects of the canon:

> Here the ornamental qualities of plants are paramount – no ecological concepts of community or association becloud the objective. Plants are analogous to domestic pets, dogs, cats, ponies, canaries, and goldfish, tolerant to man and dependent upon him; lawn grasses, hedges, flowering shrubs and trees, tractable and benign, are thus man's companions, sharing his domestication. This is the walled garden, separated from nature: a symbol of beneficence, island of delight, tranquility and introspection. It is quite consistent that the final symbol of this garden is the flower.[18]

He elaborated further on the disconnection of gardens from the surrounding environment: 'Not only is this a selected nature, decorative and tame, but the order of its array is, unlike the complexity of nature, reduced to a simple and comprehensible geometry. This is then a selected nature, simply ordered to create a symbolic reassurance of a benign and orderly world – an island within the world and separate from it . . . It is indeed only the man who believes himself apart from nature who needs such a garden.'[19]

His holistic outlook dealing with large-scale environments was also finely tuned to the other end of the ecological spectrum, with a prescient acknowledgement of the importance of soil and the life that it contains: 'therein are these essential decomposers . . . the return stroke in the cycle of matter in the universe, these unknown, named heroes, essential for all life processes . . . these decomposers reconstitute the wastes of life and the stuff of life after death, so that life may endure.'[20]

While McHarg never proposed specific planting strategies, instead focusing on plant communities as indications of environmental conditions to be read along with other signs from the landscape as constituent parts of a bigger picture, he nonetheless recognized a plant-centric perspective of ecology:

> The plant really is the basis of the creative process that is the earth, it is a little plant or the chloroplast in a plant that holds its leaves up to the sunlight and says, 'Sun, may I have some of your energy on its path to degradation, I will give it back to you but may I have it temporarily?' . . . All life, all plants, all animals, in all time . . . have depended on the dialogue between the sun and the plant, in which the plant temporarily entraps sunlight on its path to entropy, and encapsulates it into its being from which then derive all the orderings accomplished by all life in all time . . . Plants are not beautiful, they are indispensable. They may also be beautiful, but the beauty is a bonus.[21]

McHarg painted a picture of natural processes with broad ecological brushstrokes, which neglected to fully appreciate some of the finer detail and more unpredictable interconnected elements of ecosystem dynamics, preferring instead a rigid instrumental perspective in which they could be mapped, quantified and then applied mechanistically within a design framework. His rationalism was underwritten by a no-nonsense ethical sensibility that anticipated an improved future in which humans were more integrated and at ease with their surroundings, and he brooked no dissent from nonbelievers in his creed: 'I conceive of non-ecological design as either capricious, arbitrary, or idiosyncratic, and it is certainly irrelevant. Non-ecological design and planning disdains reason and emphasizes intuition. It is antiscientific by assertion.'[22] While many have since taken exception to some of his forthright views, his teaching at the University of Pennsylvania has provided an enduring legacy by laying much of the groundwork for ecologically orientated landscape architecture today.

SEASONAL CHANGE

The allure of the prairies has continued to be a strong draw for contemporary landscape architects. A restoration project at Morton Arboretum in Illinois, designed by arboretum curator Ray Schulenberg in 1962, set the bar high, transforming former farmland into a prairie as a means to address the loss of native eastern tall-grass species and habitats in the Midwest. In order to ensure that only genetically appropriate plants were used in the project, seeds were collected from railway sidings and neglected land remnants within an 80-km/50-mile radius of the arboretum. The first 3 hectares/8 acres of the 20-hectare/50-acre site were planted with seedlings raised from the collected seed, and included more than 250 species. The plants were allocated in a pre-portioned manner based upon certain percentages for each species, and then laid out in a random manner and augmented with broadcast seed. Once established, they were managed by controlled burning. Over the following decades the site expanded to 40 hectares/100 acres, and featured various prairie ecosystems as well as oak savannah.

Continuing the tradition and inspired by the work of Jensen, and Roberts and Rehmann's book *American Plants for American Gardens* on plant ecology, landscape architect Darrel Morrison has been adept at reading the landscape and interpreting it in ways that both educate and inspire people and also work for the benefit of biodiversity. He believes that tapping into the regional flora, and using it as a source for planned landscapes, can help with conservation of threatened species and create the potential for enhanced experiences of the local character, as well as requiring less resource inputs. On a larger landscape scale, local flora can create an interconnected mosaic of sites that are beneficial for wildlife and facilitate migration.

While prairie restoration has been a large focus of his work, he is nonetheless realistic that appealing to an idealized, long-lost landscape is impractical, and that it is instead necessary to create aesthetically interpreted versions that are a synthesis of art and ecology, where 'not only do you want to hew to the ecological integrity, you also want to create a space that is experientially rich; it should look beautiful, and there should also be sounds and smells. There should be other forms of life too – birds, butterflies, and other pollinators.'[23] Designed spaces for Morrison should create spatial frameworks in which native plant communities can be distilled and celebrated: 'I believe landscapes we design and manage should be experientially rich, ecologically sound, of the place, dynamic, changing over time.'[24]

While these plant communities are simplified abstractions arranged in a legible way that can be more easily understood by the public, their reduced form will always 'contain the most important species of those communities, ecologically and aesthetically, and distribution patterns which express or even heighten the unique character of those natural communities'.[25] Planting is done by plugs or by seed, employing a range of approximately forty to seventy species per half hectare/acre. Morrison's arrangements involve species flowing in rivers and drifts, with groups of one species gradually dissipating into the next one. Grasses play a central visual role in creating a sense of continuity throughout a prairie scheme, and a practical role in maintaining competitive balance and ensuring soil structure. Morrison aims to produce a complex yet coherent balance, with the colour and textural properties of the plants accented by the play of light and shadows across the community. The influence of landscape architect Arthur Edwin Bye's *Art into Landscape, Landscape into Art* (1983) is evident in the way that Morrison aims to utilize backlighting from the sun to produce translucent effects with the foliage of ferns and grasses. All of these factors lead to the creation of a sense of mystery, something that Morrison uses as an enticement to encourage exploration, an idea developed in the research of environmental psychologists Rachel and Stephen Kaplan of the University of Michigan.

Working with the old adage that the only thing constant in nature is change, Morrison rejects the traditional approach to design as a lost opportunity that misses out on the rich experience of the changing landscape over the years and decades. Successional processes are at the heart of his outlook, and initial plant communities may either contain all the desired species at the outset, such as in a prairie planting, or only harbour the first-phase species of a process, as in a forest. Setting up these scenarios and relinquishing control to variable factors are part of the intention: 'It can and should change over time within certain spatial frameworks that we establish. I'm very happy in a design I do if I see a certain amount of plant reproduction and migration change in patterns over time because the plants are telling us where they want to be.'[26] This also reduces maintenance efforts, which are then focused on eliminating any initial unwanted invasive species in order to ensure adequate density and diversity of native plants; in the long term this will act to prevent further incursion. Occasional burning is also a means to provide a reset and reinvigorate general growth.

His design for the Native Plant Garden at the University of Wisconsin Arboretum, created between 1997 and 2002, used species from eleven plant communities native to the local area, including prairie, savannah, forest and fen. The process involved removing the existing vegetation on the site, creating a groundcover layer, and establishing the communities across the 1.6-hectare/4-acre site.

At the Brooklyn Botanic Garden Native Flora Extension, conceived in 2011, Morrison's planting design strategy involved identifying zones for different plant communities, including pine barrens and plains, each hosting various moisture gradients within them (Plate 10). Plants were then selected 'based on three categories within each natural model: the characteristic *dominant* species, the most *abundant* species, and the *visual essence* species'.[27] And in the placing 'of all the plants – tree, shrubs, and ground layer species – I tried to emulate the "directional drifts" seen in nature, sometimes very loosely aggregated, sometimes closely spaced, depending on the species'.[28] Aside from putting in 15,000 plants as plugs, 30–38 cm/12–15 inches apart, native grass seed was broadcast. Opportunistic non-native plants taking advantage of patches of bare soil during establishment were removed by hand. Other notable projects include native gardens at New York Botanical Garden, Chicago Botanic Garden and the Lady Bird Johnson Wildflower Center in Texas (Plate 11).

Influenced by Aldo Leopold's essay 'Land ethic', which considered humans as an equal part of a landscape 'community, alongside everything else within it, including soil, water, plants and animals',[29] Morrison believes that landscape architects have a moral duty to increase the biodiversity on the sites they work on, and that any decrease on what might have been there previously is a failure: 'It is really our responsibility to protect the species that were here before we were; we've done a really good job of diminishing that diversity, or a bad job you may say, over these last couple hundred years, where we've taken really diverse systems, oversimplified them, made them dysfunctional and have spent all kinds of resources on chemicals and fertilizers that in the end are negative.'[30]

The design profession's role is also to highlight these issues in a way that facilitates experiential learning. By showcasing the beauty of plant communities to encourage exploration, Morrison believes that 'people will first be attracted by what's pretty –the attention grabbers like mountain laurel or lupine. Then once they're drawn in, they start to notice patterns and differences. I see it when I'm teaching field courses with students. First they'll notice the colourful flowers, then the surrounding grasses and ferns, then the greater environment and the complex ecology of the place. I love to get people "hooked" in this way.'[31] Once they are engaged, 'then this leads to starting thinking about the processes that have led to the patterns. It is a progression.'[32]

Native plant communities are also important to the work of landscape architect Larry Weaner, whose approach to planting digs deep into the ecology of meadows, woodlands and the ecotones between them. Eschewing all of the traditional principles of horticulture, he instead draws on restoration ecology translated into a method he calls 'garden ecology', which he describes in his book *Garden Revolution* (2016). Working on a large domestic scale, Weaner has had the space to stretch his ideas and develop them in practice over three decades, observing and learning from the plant communities as they progress over time in concord with the site. Planting design for Weaner is not something done prior to establishment of a garden, but rather an ongoing engagement in which the dynamic interactions between the plants and the site are not simply a part of the process but actually the desired result. More than most other ecological approaches to planting, Weaner's method embraces, emphasizes and enhances continual change as the ongoing basis of an ever-morphing garden.

Weaner's advocacy of native plants growing in a community is premised upon his belief that they both look and work well together due to their evolutionary compatibility: 'If you're taking a plant community and putting it whole, or close to whole, in a garden, you've got plants that are almost by definition going to interact gracefully because they've been doing this, and they've sorted their spatial arrangement out, for hundreds of years.'[33] Plants within a native community that are regionally and locally adapted are able to fill every available spatial niche, both above and below ground, as well as temporal niches through seasonal change. Weaner reasons that 'filling all the niches helps to exclude weeds, and so does arranging the plants synergistically. In a woodland, shrubland, or meadow where all the strata have been filled, there are so many layers of plants an invader would have to penetrate that the planting, once mature, is notably weed-resistant. Planting synergistically provides aesthetic benefits as well: the orderly, if sometimes irregular, structure of a skilfully planted woodland, shrubland, or meadow presents a harmonious, graceful picture.'[34]

Initial implementation is by either planting or seed sowing, depending on the site conditions, and Weaner's concern is in ensuring that appropriate density and niche coverage are achieved through plant selection that balances out the ruderal and competitive tendencies of the species introduced. It is important at all stages of the planting's progress: 'Annuals, biennials, and fast-growing perennials establish cover during the first year or first few years, whereas long-lived plants may not have a serious visual presence for many years.'[35] Accepting that plants play transient roles and have limited lifespans requires relinquishing the traditional notion of idealized static gardens, as well as the control that is needed to enforce them, and instead adopting flexible and reactive responses to the communities' development.

For Weaner planting design is not simply a matter of plant selection and placement, but rather a long-term engagement with a site that looks to both the past and the future. His approach begins with an in-depth site analysis: 'What are the existing environmental conditions of the property? What plants naturally inhabit those conditions? And equally important, but rarely considered, what natural processes of change will affect the landscape efforts going forward?'[36] Looking beyond just the typical factors such as soil, water and climate, he is interested in the plant communities that existed originally on the site, what forms of disturbance have affected them, and what would happen if all intervention on the site were removed. Thinking about these successional processes requires understanding how the vegetation that originally grew in a particular place may influence the plant composition in the future. For Weaner, the seed bank is the repository of the original plants that occupied the site, and therefore an indication of the plants that may naturally develop within a newly planted scheme. These pose a bigger threat to the success of a new planting than any invasive species that may appear after initial establishment. His approach draws heavily upon Frank Egler's idea of 'initial floristic composition', taking into consideration the existing seed bank in the soil and the way in which it can be directed if it materializes, which means that strategies can be developed so that the appearance of undesirable plants can be minimized, and desirable species welcomed and allowed to flourish.

For Weaner, understanding the different types of reproductive strategies of plants allows ongoing management to become a partnership with these processes. Self-seeding becomes a form of natural recruitment in which those species that are considered desirable can be kept and those that are unwanted can be removed. 'Ecology-minded gardeners often refer to the benefits of including a wide diversity of native plant species. This is certainly true, but including plants with a wide array of proliferation strategies can be equally important when self-perpetuating landscapes are the goal. Plants have different strategies for propagation, and you need to include as many of those strategies into your landscape as possible. Understanding how plants propagate also allows you to maximize the speed with which they colonize.'[37] Deploying these strategies means that 'over time the seed bank converts from a weed dominated soil strata, to a native plant dominated soil strata, which means when there are openings for a weed to get in, for whatever reason, as time goes on, there's a higher chance that it will be a native species that you're considering desirable, filling those gaps instead of a weed.'[38]

In order to shape the direction of these mixed communities of intentional and spontaneous plants, Weaner also suggests subtle forms of intervention, such as selective trimming at different levels to prevent the growth of dominant species that does not affect others. 'An overriding principle in managing for desirable and against undesirable species when they are both present in the same area is to exploit the species' differences. If you want to promote or discourage any particular plant or group of plants, identify how those plants differ from those around them and then make use of that difference.'[39] While maintenance reduces as time goes on and communities reach different forms of stability, Weaner's methods realize that gardens are an ongoing engagement with plants through understanding the ecological dynamics that propel them.

OEHME VAN SWEDEN/PHYTO

Morrison's approach to prairie planting was given a stylistic twist in the work of landscape architects James van Sweden and Wolfgang Oehme, who created a trademark identity distilling the vastness of scale and swathes of grasses and perennials so redolent of native meadows into residential and public realm settings. Also drawing inspiration from nurseryman Karl Foerster and designers Mien Ruys and Roberto Burle Marx, they synthesized these elements into a distinct landscape language, breaking abruptly from the European influences that had dominated most of garden design when they began in the 1970s. Given the name 'New American Garden' by Mark Cathy, Director of the National Arboretum, in 1980, their plant-laden designs quickly gained traction on the East Coast. Explaining their novelty in *The New American Garden* (1997), van Sweden asked, 'What's new about our New American Garden is what's new about America itself: it is vigorous and audacious, and it vividly blends the natural and cultivated. It is relaxed and rich, and it celebrates seasonal change.'[40]

Their painterly approach employed masses of grasses to create naturalistic spaces with a free-spirited feel: 'tapestry-like plantings and perennials and masses of the same plant – 3,000 black-eyed Susans instead of six'.[41] Their plantings were intended to be low maintenance and seasonal, dynamic and informal, with intermingling plants and a certain amount of seeding: 'We never really worried about colour, we worried more about texture, height, and achieving drama in the garden with plants, but not so much about colour, we just let the colours flow, and colours do their thing.'[42] When shorn of its ecological intentions, the OvS planting style was distinctly an aesthetic vision, with ornamental qualities as the driving force and art as inspiration.

As an assault on traditional garden sensibilities, lawns were removed and dense planting placed in front of the house in what was seen at the time as a revolutionary act. This thrust to change homeowners' minds about the notions of what is acceptable as a garden has been key to the work of a former OvS employee, landscape architect Thomas Rainer, and plantswoman Claudia West. Together they established Phyto Studio and co-authored *Planting in a Post-Wild World* (2015), setting out a mission statement that challenges the accepted norms of American gardens and public spaces in which isolated plants sit in a sea of mulch, devoid of either ecological or aesthetic merit.

Phyto's approach has a foundation in the German tradition of ecological planting – in particular, the tradition of mixed perennial communities developed at Hochschule Weihenstephan-Triesdorf, where West studied before relocating to the United States. Their work has sought to translate ecology's scientific underpinnings into accessible forms through their landscape designs, as well as disseminate the knowledge to other designers and the general public, by 'trying to make the most useful parts of those ideas relevant to a broader audience, and to explain why diversity is really important and functional in landscapes. Something other than traditional methods of horticulture can be beautiful, but people seem afraid to try them, so we want to bring to their attention how things could be different and why that's a good thing.'[43]

In *Planting in a Post-Wild World* they highlight the importance of understanding the key landscape archetypes such as grasslands, woodlands and shrublands, forests and edges, for the characteristics they provide spatially, seasonally and botanically. Within each of these habitats, the layering of plants plays an essential role in determining how they function and look, and is always a prime consideration for any designed planting scheme. Filling every niche and creating complete ground cover is essential in ensuring that plants work together as a community. Central to this is taking into account the forms of plant structure, foliage shape and root morphologies in order to determine which species can comfortably coexist without overcompeting.

They identify two design layers and two functional layers. First, structural framework plants to form the visual focus include trees, shrubs, upright grasses and large-leaved perennials, while mid-height seasonal theme plants provide colour and texture, often in masses or drifts complementing the former group. One functional layer features low, shade-loving stress-tolerant species that provide ground coverage, as well as plants with wide, flat basal rosettes that minimize unwanted intrusions, while the other accommodates short-lived ruderal species that fill gaps and add seasonal interest. Annuals are also employed, particularly in projects where disturbance is a highly likely regular occurrence, such as stormwater systems. On such occasions the annual species in the seed bank will be activated and will germinate to plug the gaps in a self-healing manner that reduces maintenance.

Management is a crucial factor that determines the sophistication of design in the public realm, and Rainer and West aim to ensure that the long-term requirements of a scheme are met by the skill set of those overseeing it so that it will be sustainable. Articulating the intentions behind their work with those in municipal authorities is an ongoing process, the aim being to progress the dialogue around ecological planting. Dialogue is also something they want to see broadened across all sectors of the landscape industry, with ambitions 'to raise a village between growers and designers, getting nurseries to grow more functional plants, not only flower-heavy darlings, but to embrace more functional thuggish groundcover plants that can tie plantings together and provide herbivore benefit'.[44]

On a more general level, educating the public to the benefits of ecological horticulture and getting the suburban homeowner who doesn't know what to do with their land to be more accepting of the look and aesthetic of their approach are essential. Phyto are under no illusion that it can be anything other than a hard sell to convince people to adopt and accept planting that may seem wild and untended. They believe the key responsibility for designers is to ensure that designs are attractive and legible, framed within contexts that people understand and can relate to. Distilling the characteristics of a planting scheme and amplifying and exaggerating them can be a useful strategy to make them visually coherent, by finding low, short mixes that are really emotional for people, hitting them at a gut level, and using colour as a 'gateway drug' that will get to a lot of people 'as long as you are conscious about framing it well and thinking about off-seasons, to make it palatable at its low points' (Plates 12 and 13).[45]

Balancing the needs of people and plants is the juggling act that drives Phyto and the methods they are shaping to local situations: 'We're starting to respect that deep acceptance of certain aesthetics more now, and are trying to meet half way between what plants need, diversity we need, naturalness we need, and what people naturally perceive as beautiful, and that's exactly the hybrid system that designed plant communities are showing as a solution.'[46]

GROWING WILD IN THE NETHERLANDS

HEEMPARKS

In much the same way that Robinson and the Arts and Crafts movement kicked back at the increasing industrialization of their time, a similar movement began to take shape in the Netherlands in the early twentieth century. Ecological approaches to landscape design began as a reaction to expanding urbanization in a country where the environment was a historical product of intensive manipulation and management. The new approaches developed at a scale beyond that of the domestic garden, manifesting themselves instead in the creation of a series of public parks in which a variety of naturalistic approaches to planting were explored.

In the 1920s, Dutch teacher and biologist Jacobus P. Thijsse expressed concern over the loss of semi-natural landscapes, realizing that the importance of people experiencing these environments was something that traditional parks and gardens were unable to deliver (Plate 14). When he was offered a 2-hectare/5-acre site in Bloemendaal, near Haarlem, by the local municipality for his sixtieth birthday in 1925, he set about creating an educational park featuring semi-natural vegetation. The idea was to provide an environment that would be an instructive resource for people to learn about plants and habitats. The park was eventually named Thijsse's Hof after his original wish to name it Garden of E(e)den, for the botanist Frederik van Eeden, was deemed inappropriate by his friends. It contained a variety of habitats, including a small woodland, a pond and marsh, heathland, a German landscape, and a cereal field with rare arable weeds containing native woodland, scrub and grassland plants.

The park was coordinated by Thijsse and designed by the landscape architect Leonard Springer, with construction and planting by nurseryman and landscape gardener Cees Sipkes. Sipkes was a native plant enthusiast who cultivated a select choice of plants, intended for specific ecological conditions, 'that stand out for their beauty, special biological properties, or because they are characteristic of the character of the landscape'.[1]

With a similar agenda, the educational Zuiderpark in The Hague was designed by Alida Johanna Ter Pelkwijk and created between 1933 and 1935 to provide a context to experience a variety of native plant communities on a 3.5-hectare/8.5-acre site. Each community was planted in appropriate soil media reflecting their habitat or original region, and included woodlands, indicative of the Limbourg area, and heather moors, such as those found on hillsides around Utrecht. Another early Thijsse-related venture was Amsterdam Bos, a public facility with forest and lawns designed under the direction of Cornelis van Eesteren and Jakoba Mulder in 1937. The 1,000-hectare/2,470-acre park features a diverse landscape of wet and dry areas, woodland, grassland, reedbeds and waterways, and was intended for the urban residents of Amsterdam.

The idea of naturalistic parks formed into the Heempark movement with the creation of a number of public spaces that showcased native plants. The term (to be translated as 'habitat park') was introduced by landscape architect Christiaan Broerse, whose vision was to present in one place a series of different naturalistic landscape typologies, each distilled into a readable form through the combination of key indicative plants. This method balanced the aesthetic and ecological aspects of the planting in order to make them appealing to the public rather than simply educational as the earlier parks had been, showing that native planting could be colourful and attractive. This entailed creating an experience of Dutch landscape typologies that were simplified ecologies – featuring herbaceous perennial plants with shrubs and trees, and based upon plants' needs – rather than accurate facsimiles of natural plant communities.

Broerse was Director of Parks for Amstelveen, an expanding area on the southern outskirts of Amsterdam, and working with botanist Jacobus (Koos) Landwehr instigated the most well-known heempark, Jac. P. Thijssepark, in 1939. Established on 2 hectares/5 acres of former agricultural land as a series of meadow, heath, woodland, dune and bog, habitats were created reflective of the local acidic peat substrate. The availability of differing moisture conditions, due to the varying proximities of the ground surface to the water table, allowed for a diversity of plant communities with specific niches within them and gradients between them. Peripheral woodland planting facilitated screening out the surrounding built environment (Plate 15).

The controlled aspect of the park meant that its development was very much an ongoing interaction between the plants and the staff, who were charged with ensuring plants were allowed to develop in their own direction while avoiding unwanted degrees of dominance and succession. A highly skilled maintenance regime was installed in order to ensure that the park had a dynamic sense, with different states of carefully managed succession. The delicate nature of the peat substrate meant that mechanical intervention was not possible, and instead intensive weeding by hand was needed to prevent competitive and unwanted invasive species from upsetting the desired balance within the plant communities. Seeds were harvested on site for propagation and future planting in order to maintain a local genetic integrity and to avoid plundering the seed supply in semi-natural areas. The management of the park by Hein Koningen over succeeding decades has ensured an ongoing fidelity to the original intentions.

Two decades after his experience in the creation of Thijsse's Hof, Sipkes was involved in another project, Heempark De Tenellaplas at Voornes Duin, near Rockanje, created in 1949. Named after the rare bog pimpernel *Anagallis tenella*, the 3.5-hectare/8.5-acre instructive park featured a unique coastal dune ecology with lakes, creating a wet to dry moisture gradient, as well as heathland, providing a diversity of three hundred species. Sipkes's book *Wildeplantentuinen* (Wild Plant Gardens), written together with Landwehr in 1973, pooled their extensive knowledge to provide a guide with practical plant lists for different biotopes, including perennials, bulbs and tubers.

SPONTANEOUS VEGETATION STRATEGIES

In the 1960s artist and gardener Louis Le Roy developed an approach that preceded later ideas about urban ecology in seeking to integrate people within their surrounding landscapes. Critical of humans controlling the rest of nature, Le Roy believed that they should instead creatively collaborate with natural processes in order to achieve balanced relationships with vegetation, soil and animals. He saw a large part of the problem to be the dissociation of local residents within the developing towns from the municipal structures of authority that shaped their environments through design and planning processes. Challenging these hierarchies, he encouraged community participation in the creation and management of planted landscapes. The principle of self-organization was the central link between the ecological and social aspects of Le Roy's world view. Yet despite his ambitions, his rather haphazard approach to appreciating environmental conditions and plant requirements meant that his endeavours often met with less than successful ecological results.

His first project, a 'people's park' with wild planting, the Kennedylaan in Heerenveen, was a strip of neglected grassland sited between two housing developments. Into this strip, which was 18 metres/60 feet wide and ran for 1.5 km/1 mile, from woodland at its outer reaches to the town at the other, he introduced over a thousand species of both native and introduced trees, shrubs, perennials, ferns and bulbs. Le Roy recognized that traditional landscape and garden design was premised upon a static vision in which natural processes were controlled, and advocated instead a dynamic approach to planting. His strategy was to let plants fend for themselves and self-organize, with an absolute minimum of human intervention, a view inspired by spontaneous plant communities that he had observed in a disused local canal. He anticipated that the plantings would stabilize in diverse, complex and visually appealing climax states over a period of years, in line with the ecological ideas of American biologist Eugene P. Odum, but Le Roy's misconceptions about successional processes led to other results. The competitive tendencies of plants such as nettles (*Urtica dioica*) and sow thistle (*Sonchus oleraceus*) soon made their presence felt, to the exclusion of other plants. Likewise, a lack of understanding of the basic soil conditions resulted in the loss of many trees that had been planted into inappropriate media.

More successful, however, was the social uptake of the parks and other projects that he initiated in Leeuwarden, Eindhoven and Lewenborg, a housing suburb of Goningen, where in 1972 he undertook a ten-year contract on a 6-hectare/15-acre site to develop gardens in collaboration with the local community. Finding a resonance with people seeking respite from the dense urban areas they lived in, the ideas espoused in his book *Natuur uitschakelen, natuur inschakelen* (Turn Off Nature, Turn On Nature), published in 1973, found favour for a period on a national level.

A particularly interesting enterprise took place in Delft on the Gillis Estate beginning in 1968. Two new identical housing developments were used as an experiment to capture residents' reactions to their environments. Under the direction of Henk Bos, Director of Parks, they were both landscaped by his department and monitored by the Netherlands Institute for Preventive Medicine in Leiden. The Handellaan was laid out in a traditional manner privileging buildings and cars, with a small formal garden and courtyard, while the Haydnlaan attempted to create the richness of the countryside by introducing woodland glades into the courtyards between the buildings and making connections to the surrounding parkland.

At the Haydnlaan the spirit of Le Roy's ideas of cities was very much in evidence. Existing typological features such as ditches and flooded areas were retained to provide the potential of habitat diversity. The landscapers were also given a certain freedom to landform as they wished rather than follow a prescribed plan. Predefined planting was largely eschewed, with the existing ruderal herbaceous vegetation supplemented with three mixes of trees and shrubs of later successional stage species, with alder and rowan used as linking species, giving the appearance of scrubland reverting to woodland. Both native and introduced plants were deployed in groups according to their habitat requirements, and spontaneously appearing plants were then allowed to join them and take their course. Open areas were initially broadcast with agricultural seeds and field species, which were then colonized by adventitious competitors such as nettles, dock and thistles. Due to a local regulation these areas had to be mown, but herbicides were forbidden.

The results of the study exploring the social effects of the two housing developments revealed that the Handellaan landscape resulted in a disconnect between the physical and social environment, while the Haydnlann facilitated unstructured play and exploration, providing an informal educational resource for children, which was reinforced in the local school where gardening activities took place. The researchers recommended the lessons learnt from the Hadynlaan landscape be integrated into developments in other towns.

Attempting to avoid the pitfalls of the concentrated management of the heemparks and the wild abandon of Le Roy yet pursuing their positive points, ecologist Ger Londo, working for the National Institute for Nature Management in Leersum, developed a different approach. While in favour of natural colonization of areas by native plants, he was not averse to human intervention before plant establishment or maintenance afterwards. Critical of Le Roy for the lack of scientific grounding in his approach to spontaneous vegetation, Londo undertook rigorous research on soil composition and moisture content to analyse the distinct conditions necessary for specific plants. Based upon this information, site preparation involved creating a variety of balances and gradients within the planting media in order to encourage the maximum diversity of species, a method that could be considered an alternative to traditional plant selection. The diversity and patterning of plant communities were then left to natural ecological processes. Londo explained, 'Such gardens are called natural gardens. The method requires a thorough ecological knowledge of the species and plant communities that one wants to create from the (varied) starting environments in the garden. But once this is done properly, maintenance (mowing and occasional felling) is much easier and cheaper than that of the wild botanical garden, which always requires planting, sowing and selective weeding.'[2]

Londo laid out his ideas in 1973 in the journal *Nature and Landscape* in an article entitled 'Toward more nature in gardens and parks', where he described an ecosystem approach to creating balances between plants, animals, soil, climate and human influences. His book *Natuurtuinen en parken: Aanleg en onderhoud* (Nature Gardens and Parks), published in 1977, was an ecological manual for the creation of nature gardens, and explained the environmental preparation suitable for natural vegetation: 'Species-rich vegetations always appear to be bound to certain gradient situations, to transition zones between sand and clay, wet and dry, high and low, calcareous and limestone. So you have to provide all kinds of gradual transition zones.'[3]

Where time necessitates faster establishment than that provided by spontaneous plants, Londo advocated sowing seeds with a local provenance, in order to ensure compatibility with the site conditions: 'If the environment is suitable, the establishment of species can be greatly favoured by sowing, including that of rare plants. Without sowing, you often only get general plants in the first year. Sowing also gives endangered and rare species a chance. Some of it will disappear, but what remains is important. If the spontaneous establishment of rare species is slow, it often starts with one or a few specimens that later form a large population.'[4] However, he noted that this tended to produce populations with a narrower genetic base than would be produced by spontaneous means, which was a concern as preserving as much biodiversity as possible, including variation within the species, was part of his overall ethos.

Maintenance in a natural garden based on Londo's ideas was intended to be considerably less labour- and cost-intensive than heemparks, with only one or two mowings per year of meadow areas and occasional woodland trimming and thinning, eliminating any fertilizing, weeding and replanting.

In the 1970s, the work of Professor Pieter Zonderwijk at Wageningen University, on vegetation management of roadside verges, railway banks and watercourses, led to changes in the way that they were maintained. His aim was to reintroduce places of wild spontaneity into these marginal areas of the built environment. In the case of verges, rather than intensive lawn-style maintenance, he suggested twice-yearly mowing regimes, with biomass removed from site in order to minimize soil fertility, resulting in greater diversity in the flora and fauna. His 1979 book, *De Bonte Berm* (The Bountiful Berm), highlighted the importance of these spaces, and his work for the Ministry of Agriculture saw them implemented.

THE DUTCH WAVE: 'NEW PERENNIAL' PLANTING

In the 1980s, a set of geographically disparate but ideologically like-minded individuals from backgrounds as diverse as horticulture, art and philosophy began meeting to discuss ideas and to swap plants and their stories about them. The group featured Ton ter Linden, Henk Gerritsen, Rob Leopold, and Piet and Anya Oudolf, along with an ever-evolving cast of characters and visiting fellow travellers from abroad. The various players had in their own ways been searching for a new utopian ruralism in a romantic reaction to the heavily manufactured and managed landscapes of the Netherlands, which was expressed in various ways in their work. The burgeoning group, dubbed 'The Dutch Wave' by Swedish botanist Rune Bengtsson, fused naturalistic approaches to planting with a modern design sensibility, and was successfully exported internationally under the rubric of the New Perennial movement.

The innovations in planting design pioneered by the group were rooted in a couple of key Dutch cultural developments: the heemparks of Amstelveen and the work of garden designer Mien Ruys, whose oblique sense of geometric design, editorship of the magazine *Onze Eigen Tuin* (Our Own Garden), and personal garden Tuinen Mien Ruys at Dedemsvaart defined a decidedly Dutch interpretation of modernism. While drawing heavily on modernism's sense of space articulated through clean geometric lines, she nonetheless distinctly broke from the North American or British interpretations by featuring an enthusiastic abundance of planting, offsetting the minimal hardscaping, almost like an update on the architectural/horticultural partnership of Lutyens and Jekyll. In particular, her axiom of 'wild planting with a strong design' resonated with the group's inquisitive and experimental outlook.

Their rigorous research through growing plants and exploring them in their natural settings resulted in the articulation of a strong environmental sensibility as evidenced in the philosophical writings of Leopold and the horticultural musings of Gerritsen. Leopold's ideas were outlined in a manifesto format in *Nature & Garden Art*, published in 1995, which sought a new direction for integrating a wider assortment of plants within a garden context. Central to the group's thinking was a rejection of traditional approaches to gardens entailing continuous labour and ongoing replacement of plants in order to preserve an idealized notion. As Oudolf reflected:

> Our philosophy was: getting rid of the dogmas, of the dictatorship of traditional horticulture. We were constantly discussing this theme, as one of the central subjects in our conversations. This did not mean we didn't see the importance of craftsmanship: of course we realized that before changing things you have to know how it is actually done. Finding out all the details was quite a job. How, for instance, could you find plants that would fit the new images you had in mind? And we did find ways; one of them was dropping recurring elements into planting schemes, and using annuals to create a certain spontaneity and looseness.[5]

The group investigated the use of wild plants in garden situations and drew heavily upon the work of German plantsmen Karl Foerster and Ernst Pagels, which focused on the importance of incorporating ornamental grasses in mixed plantings to provide a looser feel as well as structure and punctuation.

The nurseries and seed catalogues they established were as much for exploring these relationships as they were for commercial purposes. Almost like modern-day plant hunters and collectors, they gathered source materials unavailable elsewhere at the time in the country, which enabled them to expand their planting palettes and develop the distinct horticultural design languages for which they have become renowned. Very much in the spirit of the heemparks and other Dutch naturalistic pioneering, they embedded collecting seed with a conservationist ethos, saving disappearing species, as well as a general sense of wonder for plants. Leopold and Dick van der Burg established Cruydt-Hoeck nursery in Nijeberkoop in 1978 to offer wild plant seeds and flower meadow mixtures, which Leopold lyrically described in their catalogue 'Big Seedlist'. The creation of the Oudolfs' nursery at Hummelo was somewhat more pragmatic, addressing the paucity of more unusual perennial plants needed by Piet for his design work, in particular prairie plants from the United States, which his book, written with Gerritsen, *Méer Droomplanten* (1999), translated into English as *Dream Plants for the Natural Garden*, helped to popularize in Europe.

The social idealism inherent in the Dutch Wave's initial ferment petered out after the death of Leopold in 2005 and then Gerritsen in 2008, yet the aesthetic influence of their ideas has endured. With Piet Oudolf as the foremost remaining flag bearer, it has been placed centre stage on an international level. Oudolf's career trajectory within the garden world and beyond has been unprecedented in recent times, due to a singular vision, determination and a business acumen that set him slightly apart from his colleagues; as garden historian and magazine editor Leo den Dulk commented, 'Piet wasn't a hippy.'[6] He has become nothing short of the horticultural world's version of a rock star, with a portfolio of works including projects such as Trentham near Stoke on Trent, the Lurie Garden in Chicago, and Battery Park and the High Line in New York. The High Line, in particular, has escalated the conversation about planting in public places to new heights in a relatively short period of time; attempts to replicate the success of the park, constructed on an elevated stretch of former railway traversing the streets of Manhattan, have popped up across the globe.

Developing his approach from observing plants in the wild, he has sought to create something that captures this naturalistic spirit – providing the sensation of nature – rather than simply imitating it in the manner of some of his national predecessors: 'It may look wild, but it shouldn't be wild. This is what you'd like to see in nature . . . a landscape you would dream of but never get in the wild.'[7] Disregarding the ecological principles of plant communities, his work is instead aesthetically orientated: 'Plants were characters I could compose with and put on stage, and that is what I do . . . I let them perform' (Plate 16).[8]

While he relies on flower colour for its impact to a large extent, he distances himself from traditional mixed border planting, using a balance that places a strong emphasis on form and texture. His book written with Noel Kingsbury, *Designing with Plants,* boldly declared his rationale: 'Traditionally a plant's colour was its most important characteristic, but . . . plants will be examined in a different light for other qualities: firstly for the shapes of their flower and seed heads; then for their leaf shape and texture; and only then for their colour.'[9] The seasonal process of change was equally important, with clear structural definition apparent through every season: 'A good structure plant will continue to look good after flowering and preferably well into the winter. This does however bias plantings heavily toward the later part of the year, as early flowering perennials either lack much structure . . . or only have weak and unimpressive seed heads . . . The definition of long-term structure as the biggest single compliment a designer can pay a plant has really changed our perennial garden flora.'[10]

A bold break from traditional horticulture, initiated by Gerritsen but popularized by Oudolf, involved allowing plants to retain their structural integrity throughout their dormancy over autumn and winter, before being cut and removed for the new growth in spring. This acceptance of senescence extended Oudolf's notion of colour in the garden: 'The way they teach gardening, it's all about flowers. And colour was one of the most important things. Then I started to look at plants that had the real character, plants that had great seed heads, plants that had, for instance . . . the grey, the brown, the blondes. They became as important as all the other colours in gardening.'[11]

Oudolf's earlier work involved large plantings of perennials and grasses arranged in blocks to create drifts of colour along a border with a naturalistic appearance;

the structure provided them with a considered sense of formality that looked consciously designed rather than semi-natural (Plate 17). These arrangements featured simple, readable mixes of relatively small numbers of species over a large area: 'Block planting says that a group is next to another group to make a beautiful combination, also that combinations should have a nice sense of repetition through the whole area so it doesn't look really rigid.'[12] These are then interspersed with scatter plants to introduce a sense of spontaneity and ensure there is not too much uniformity: 'The little individuals that loosen up the planting and unite it over the whole area, and that make your eye move through the planting, so you lose the idea of looking at blocks . . . it gives the idea of coherence and flow.'[13]

Careful plant selection, through experience of using them over many years, has seen Oudolf reduce the palette by eliminating the plants that have failed to deliver, resulting in a reliable toolkit. His more recent work has moved towards a less obviously defined and more naturalistic mingled weave of plants. The former nursery site at Hummelo provided an experimental test bed for this method: a very informal dense knit of grasses with native and introduced perennials.

While Oudolf was busy creating what was to become a new international language of planting design, Henk Gerritsen was digging somewhat deeper – exploring the ecological interactions of plants and developing a much wilder style in his garden Priona in Schuinesloot, created with his partner, Anton Schlepers (Plate 18 and 19). His work pushed beyond the idea of a garden simply looking naturalistic to a method based on ecological principles developed over years of excursions across Europe observing and researching plants in semi-natural communities. Gerritsen's phytosocial knowledge and experience of plant communities, combined with the ethos of his Dutch precursors, placed him in a position to translate ecology into the context of the garden, and to create a unique approach to such a synthesis expressed in his written work and realized at Priona.

Critical of traditional gardening's reliance on fertilizers and pesticides in its attempt to maintain ecologically unrealistic environments, he proposed instead 'natural gardening', which recognized the ecological amplitudes plants can exist within as the basis for fitting them to places in the garden. Gerritsen's *Essay on Gardening* (2008) is both a paean to plants and their habitats and a guidebook for those wishing to walk on the wilder side. In a similar manner to Grime's CSR analysis, Gerritsen considered the relationship of plants to site in terms of dynamics, by considering them as vegetation typologies suited to high, medium and low dynamic situations, and differing levels of light and moisture availability, discussing grasslands, woodlands and other plant communities. Recognizing the often-impossible task of keeping unwanted plants at bay, he sought creative solutions for accommodating plants such as ground elder, and accepted them as part of an ecological awareness, along with embracing the importance of insects and animals to plant communities. Objecting to most gardeners as aspiring either to the yields of farmers or to merely decorative results, he called for a reassessment of the idea of beauty in gardens: 'For me, however, the intrinsic value of nature has always been the main priority . . . over time, my notion of what is beautiful about the garden has changed. By acknowledging that "weeds" play an indispensable part in the restoration and maintenance of the natural order, it has become easier to appreciate their objective beauty.'[14]

PERENNIAL PRECISION IN GERMANY

GARDEN ART

In a similar manner to Britain, gardens in Germany prior to the nineteenth century had followed familiar trends of productive growing and formal design. Flower gardens featured since the sixteenth century as either cottage-style gardens with native plants or laid out in raised beds, geometric parterres or elaborate knot gardens, and often presented massed monocultures or individual plants displayed with spaces separating them. In the early nineteenth century, the English landscape garden and the Picturesque styles were touted as novel approaches to design, yet failed to gain widespread traction.

However, by the middle of the century and prior to the publication of William Robinson's *The Wild Garden*, ideas about naturalistic gardens were beginning to develop. German landscape architect and garden historian Gustav Meyer picked up on Humboldt's ideas of phytogeography in his 1873 book, *Lehrbuch der schönen Gartenkunst* (Textbook of Beautiful Garden Art), and employed them in the design of the public park Humboldthain in Berlin, with an international array of plants arranged in groupings according to their national origins. Hermann Jäger, a botanist, garden writer and associate editor of the journal *Gartenflora*, was interested – in a similar manner to Robinson – in imitating nature by naturalizing introduced plants with natives in appropriate settings such as woodlands, shrubberies and lawns. His suggestion in *Lehrbuch der Gartenkunst* (Textbook of Garden Art, 1877) to introduce additional colour into meadows highlighted the aesthetic ambitions of his ideas, although he was keen to enhance the wilder qualities of the schemes. A writer using the pseudonym Dendrophilus wrote a series of articles in the publication *Garten Zeitung* in 1882 in praise of the idea of 'The Wild Garden', a term he acknowledged adopting from Robinson: 'I found it in the English magazine *The Garden* and it struck me as apt.'[1] His encounters with plants in their habitats on his travels led him to the judgement that 'garden art is superfluous . . . these waysides, untouched by the gardener's hand, are often delightfully beautiful and much lovelier than horticultural perfection will ever be achieved, of this I am well aware.'[2] Like Robinson, he was not averse to using naturalized introduced plants familiar to gardeners at the time, and his articles listed his recommendations for preferred locations. Extolling the virtues of this new way of gardening, he invited his readers: 'Walk with me through the wild garden, look at it carefully, grasp its meaning, grasp it with your heart, and you will find in the future it is your favourite walk at this edge of the forest!'[3]

THE SOCIOLOGY OF PLANTS AND PLACE

At the beginning of the twentieth century, garden architect Willy Lange proposed the idea of a biological aesthetic for planting: grouping plants according to their appearance as they occur in natural situations in their regional habitats. For Lange, the garden was a form of immediate and intimate mediation with nature and an inseparable part of the larger surrounding landscape. In his notion of the 'Nature Garden', he recognized the importance of Ernst Haeckel's work on ecology and Alexander von Humboldt's ideas on plant physiognomy as key to ensuring that appropriate planting conditions are observed, based on reference to their natural environments. His book *Garden Design for Modern Times*, published in 1912, proposed that the Nature Garden should avoid architectural and geometric designs in favour of informality in order to relate to the surrounding environment. Adopting a non-anthropocentric approach, he embraced the energy and forms of nature as paramount and rejected human imposition on plants. He considered shaping and bending plants purely for aesthetic pleasure unacceptable, as he believed that plants should on the whole be left to establish and then fend for themselves.

While it was important to ensure that suitable ecological requirements were observed within a planting scheme, for Lange the choices of the designer were key to moving beyond a garden's being merely a facsimile of nature to become a work of art. Lange's intention was not to imitate nature but to improve upon it, and while a preference for native plants was observed, it was not a restricting factor for him, with introduced plants used as long as they suited the conditions and looked right within the plant community. An example of a 'vegetation picture' in his own garden in Wannsee, a suburb of Berlin, featured an unlikely combination of azalea species and *Sedum spurium* growing as an understorey layer beneath *Pinus silvestris*, arranged to present a visual analogy of an evergreen forest (Plate 20).

Though Lange's approach accepted introduced plants, foreign garden styles carried nationalistic connotations for him. His belief was that there were different styles for different nations, which existed within a predetermined moral hierarchy. His disdain for geometry and formalism was not simply an appreciation of natural forms, but also part of his view that these modes of expression were straitjacketing strategies used by the cultures of southern Europe to limit the power of the Nordic race.

Unfortunately, the relationship between plants and place has been not only an ecological issue but also one that has often spilled out into the political realm. Evolutionary adaptation has linked plants to places, but adding people into the mix has sometimes taken dark and dangerous turns by linking ecological relationships with ideas of racial superiority. The ideological implications within Lange's work, and the associations between German plants and people, fed into later National Socialist ideas about racial purity and the bonds between land and citizenship.

Garden architect Alwin Seifert extended many of Lange's ideas, stridently asserting the relationship between garden, landscape and people as forming a natural biological harmony. His particular take on the Nature Garden was to discount introduced plants in favour of natives. Seifert appealed to the notion of *bodenstiindig* (variously interpreted as indigenous, down to earth and rooted in the soil) and in the magazine *Die Gartenkunst* in 1929 he proposed, 'As commonly accepted, a down-to-earth garden would be one that would be covered with native plants preferably and that in its underlying mindset would be affiliated with an approach to garden design that one would understand as particularly German.'[4] Slightly later, his stance softened a little; he conceded a certain degree of latitude for introduced plants that were in harmony with their surroundings in private domestic spaces, but they were still to be strictly forbidden in public places, and he remained adamant about the use of natives in the wider landscape. Seifert's ideas were enthusiastically adopted by the National Socialists, who appointed him as the Reich's 'attorney of the landscape' to ensure that appropriate plant species adapted to regional soil and climatic conditions were chosen for the sides of autobahns in 1936.

Seifert's work on roadside planting was based upon empirical field research by botanist Reinhold Tüxen, who established the science of plant sociology in Germany using the methods developed by Swiss botanist Josias Braun-Blanquet, which determined plant community structures based on floristic composition rather than biogeographical or physiognomical criteria. A commission to map the vegetation of the province of Hanover in 1933 provided Tüxen with an opportunity to put the science into practice and create the basis for regional plant studies across the country. His 1937 book, *Die Pflanzengesellschaften Nordwestdeutschlands* (The Plant Societies of Northwest Germany), became a standard reference for other plant sociologists, and his notion of 'Potential Natural Vegetation', introduced in 1956, brought a new aspect to climax theories, incorporating the existing effects of humans on the vegetation communities as a factor which, along with the natural site conditions of soil and climate, would influence their anticipated eventual stable states in the future.

Alongside the developments in plant science and landscape architecture, new approaches to planting evolved within the field of experimental plant breeding. Inspired by Lange's work yet opposed to its inherent nationalism, nurseryman and keen perennial breeder Karl Foerster pushed perennial planting further towards more naturalistic directions by creating a planting style using hardy plants from around the world suited to local soil conditions and climate. In the nursery and garden at Potsdam-Bornim, near Berlin, which he managed from 1910 until 1970, he bred a wide selection of hardy perennials from around the world with the aim of improving growth habitats and resistance to pathogens, as well as enhancing hardy characteristics such as drought and heat resistance and longevity. While he introduced a number of wild plants to gardeners, he was also responsible for breeding around 370 crosses, with a specific focus on delphiniums, phlox and grasses, the last having an enduring effect on naturalistic garden design.

Foerster was a prolific writer, and his extensive knowledge of plants was passed on to the public through his many books. His 1911 publication *Winterharte Blütenstauden und Sträucher der Neuzeit* (Hardy Flowering Herbaceous Plants and Shrubs of Modern Times) introduced new ideas about using hardy plants, while other books, such as *Lebende Gartentabellen* (Living Plant Schedules, 1940) and *Neuer Glanz des Gartenjahres* (New Splendour of the Garden Year, 1954), provided extensive plant lists and guides to their seasonal characteristics. The dissemination of his ideas was reinforced through the garden design practice Gartengestaltung Bornim, which he formed together with landscape architects Hermann Mattern and Herta Hammerbacher in 1927 and which lasted until 1948. The practice benefited from the growth of new middle-class garden owners and a reactionary response to the growing industrialized cities and the environmental degradation that accompanied them, creating a market for more naturalistic gardens.

Facing the reality of the modern world, the self-explanatory title of his 1925 article, 'Blumengarten für intelligente Faule' (Flower gardens for intelligent lazy people), recognized the requirements of lower-maintenance gardens for home owners: 'The garden as a work of art only seems to me perfect when its maintenance work . . . adheres to very specific limits.'[5] Foerster proposed that the key to success was appropriate plant selection, including the use of sturdy and strong perennials that did not require staking, ground cover to reduce weeds, substituting suitable perennials for lawns in shady or dry areas, and appreciating the natural form of shrubs instead of pruning them.

His vision – integrating native and introduced, wild and cultivated species – was of a wild herbaceous gardening for which natural vegetation such as woods, water margins and steppe provided reference points, and aimed at creating a distinctly new type of garden style rather than slavish replication: 'Gardens should increasingly reflect the natural world and seek for new ways to develop and re-create it.'[6] Former employee Ernst Pagels followed in Foerster's footsteps, and introduced new varieties of perennials that also proved popular with later devotees of naturalistic planting.

Fusing the different approaches of his precursors, Richard Hansen drew on his experience working for Foerster in his nursery and as Tüxen's research assistant at the central office for vegetation mapping in Stolzenau. His knowledge of plant sociology led him to develop a methodology based upon the ecological requirements of different species when combined in communities within different site conditions in the garden. Plant relationships and habitat typologies influenced the groupings but, like Foerster, he was keen to integrate cultivated and introduced plants alongside natives. Instead of following the traditional associations of plant sociology (*Pflanzgemeinschaften*), Hansen's plant communities were consciously designed and related to different environmental conditions found in gardens rather than replicas of wild communities. His systematic approach established a series of garden habitats with associated plant lists, and featured species that were most appropriately suited ecologically to endure in relatively stable, long-term communities.

Hansen was appointed head of the Institut für Stauden, Gehölze und angewandte Pflanzensociologie (Institute for perennials, shrubs and applied plant sociology) in Weihenstephan in 1948; here, encouraged by Foerster's work, he established the Sichtungsgarten, or viewing garden, to conduct long-term plant trials researching different garden habitats. The research carried out by Hansen and Hermann Müssel aimed to investigate favourable growing conditions for perennial plants in the garden based on their natural habitat requirements as well as according to their aesthetic qualities. This habitat-based concept of 'Lebensbereiche' (literally translated as living spaces) assessed species and cultivars in terms of their growth rates and habits, life strategies, reproductive tendencies and maintenance requirements when arranged together. Intracommunity interactions were key to the methodology, ensuring that while both native and introduced species had similar origins, they actually proved to be compatible when placed in new combinations in communities. The outcome was to produce an extended flowering season in visually harmonious communities requiring minimal maintenance. A degree of annual variation was permitted through plant reproduction without compromising the overall integrity. Intrusions of unintended plants were generally tolerated as long as they did not exhibit overly competitive traits, their presence minimized by the informality of the arrangements.

GARDEN HABITATS

The outcome of the research by Hansen and Müssel resulted in an index system and lists of perennials published in 1972 and then in a 1981 book co-authored by Hansen and Friedrich Stahl, *Die Stauden und ihre Lebensbereiche* (Perennials and their Garden Habitats), which proposed a series of seven distinct garden habitats that relate to natural settings for plants translated into a designed garden setting. These habitat typologies were woodland, woodland edge, open ground, rock garden, bed, water's edge and water. A further subdivision according to environmental conditions accounts for different local conditions. Moisture availability was broken down into dry soil, fresh soil, moist soil, wet soil, shallow water, floating-leaf plants, submerged plants and floating plants. Light availability was characterized as sunny, indirect light, partially shaded and shady. Sociability factors were then attributed to plants based on their tendencies to grow individually or in groups in order to determine the number of plants allocated throughout a design, ensuring that natural competitive traits would be balanced out within the communities. These were to be planted individually or in clusters, in small groups of 3–10 plants, large groups of 10–20 plants, large groups over large areas, and intensively planted over large areas. Cross-referencing all of these factors produced an abbreviated identification code that could be related to a list of appropriate habitats. This list was later expanded by perennial specialist Josef Sieber and adopted by nurseries as a standard reference system for categorizing plants.

While the detailed complexity of the system created something of a barrier to widespread uptake of the system, the influence of Hansen's method continued in the work of Rosemarie Weisse, who created the planting scheme in Munich's Westpark, and that of Hansen's students Hans Simon and Urs Walser. Hans Simon's experimental work with perennials, as well as drought-tolerant shrubs and trees, at his nursery and garden in Marktheidenfeld gained some renown; landscape architect Walser's work included Killesberg Park in Stuttgart.

Walser was also involved in setting up and running the botanical garden at Hermannshof, a former large family house in the centre of Weinheim where the traditional grounds were transformed into a show and trial garden based upon a suggestion by Hansen. It was laid out in 1983 by landscape architect Hans Luz with planting plans by Walser, who assumed directorship of the non-profit organization.

The layout of the garden includes a series of stylized, dynamic plant communities for garden and urban environments based on Hansen's Lebensbereiche system, including woodland, woodland edge, steppe, dry open ground, moist open ground, water edge, water and bed. These communities have been inspired by particular global communities, such as North American tall-grass prairie, Central Asian steppe, Mediterranean scrubland, European and North American moist meadows, and East Asian temperate forest, but given the experimental orientation of the garden, the communities also feature plants from other regions that share similar requirements. The research mandate has been to observe the relationships between plants under various cultivated conditions, with a particular focus on their competitive tendencies, to develop aesthetically pleasing combinations with respect to form and colour, and to monitor permanent plant communities and their management strategies (Plate 21).

Since 1988 the directorship of the garden has been under landscape architect and horticulturist Cassian Schmidt. As well as investigating novel plant associations, he has increasingly focused upon developing time-saving management systems that are more effective and less labour-intensive than traditional horticultural methods. He believes that providing evidence-based information will assist in convincing local authorities of the benefits of planting these habitat-based schemes, and encourage them to adopt them in public spaces. Logging the time the garden staff spend on tasks relating to the upkeep of each area of planting, and keeping records over successive years, provide convincing figures that reveal his methods outperform conventional ones in terms of financial and resource inputs.

Management methods are considered in terms of Grime's CSR analysis. Competitive plants in sunny or slightly shaded locations, prairie, meadow, wetland and woodland edge communities have proven to be low to moderate time input, 7–12 minutes per year per sq m/sq yd, involving cutting back and mowing the previous year's biomass in late winter, and also remnants of the early summer-blooming vegetation in summer; these are shredded and used as a mulch on the rest of the planting. For stress-tolerant planting, such as steppe-style or dry habitats, maintenance is low to very low, 5–7 minutes per year per sq m/sq yd, with dried remains strimmed and removed. Drawing on this research, management schedules highlighting timelines, with instructions regarding the work to be carried out at each point, have been produced.

But the main impediment to uptake of the methods pioneered at Hermannshof has been that, while labour time is minimized, skilled workers are usually needed to carry out the work, as only they can identify and understand the specific habitat communities; therefore, the battle for Schmidt has been to educate municipal authorities to upskill and retrain their workforces.

Schmidt's practical excursions in the public realm, along with work by his wife, landscape architect Bettina Jaugstetter, have helped to disseminate the methods developed at Hermannshof, with acclaimed planting designs at Sheridan Park in Augsburg and at the corporate site of ABB (Asea Brown Boveri Ltd) in Ladenburg.

MIXED PERENNIAL PLANTINGS

Hermannshof has also played a role in developing a simplified version of the Lebensbereiche-style plant communities more appropriate to both the public realm and the domestic domain. The concept of Staudenmischpflanzung (mixed perennial plantings) was introduced by Wolfram Kircher and Walter Korb in 1994 at the Landesanstalt für Weinbau und Gartenbau (LWG) institute in Veitshöchheim.

PLATE 28 The carefully considered perennial meadows at Burgess Park, designed by James Hitchmough, stand up to the demands of a heavily used public space. The west-facing slopes feature drought tolerant species with moisture tolerant species in a drainage swale at the bottom, while North American prairie species populate the southern edge of the park.

PLATE 29 James Hitchmough's planting at RHS Wisley features North American prairie with western North American and Eurasian steppe species.

PLATE 30 The Dry Meadow at Cambridge University Botanic Garden mixes an international array of drought-tolerant species from steppe, prairie and grassy plain habitats with temperate species tolerant of summer drought.

PLATE 31 Nigel Dunnett's high-profile urban planting design at the Barbican in London, based upon steppe and woodland habitats, creates a seasonally changing and texturally layered garden amidst the iconic architecture.

PLATE 32 Île Derborence at Parc Matisse in Lille features a mixture of European boreal biome species and Asian and American plants. The plants have been left to create their own balanced ecosystem, visited only twice a year by the head gardener to analyse its development.

PLATE 33 (above) Gilles Clément's Garden in Movement at Parc André Citroën in Paris exemplifies one of the key components of his garden philosophy: deferring to the nomadic tendencies of plants to self-seed, colonize and move around a site, rather than imposing a static design through traditional maintenance.

PLATE 34 (left and below) Self-seeding plants including *Stipa tenuissima*, *Euphorbia characias* subsp. *wulfenii* and *Centranthus ruber* create the dynamic landscape of the Garden in Movement on the roof of a former dock in Saint-Nazaire.

PLATE 35 Accepting aridity as the baseline for planting, Ossart and Maurières insert planting into the existing landscape, blurring the boundaries between the cultivated and the spontaneous.

PLATE 36 (left) The traits of plants and communities in the garrigue are the evolutionary result of environmental conditions and cultural influences.

PLATE 37 (below) Olivier Filippi's garden is influenced by the landscape of the garrigue, with particular attention given to the forms of plants and their spatial arrangements.

PLATE 38 James and Helen Basson eschew planting plans in favour of plant mixes featuring percentages of different species based on their naturally occurring ratios in the landscape.

PLATE 39 Planting mixes are translated into graphic form, suggesting patterns that are found in the landscape, while others are scattered on a random basis.

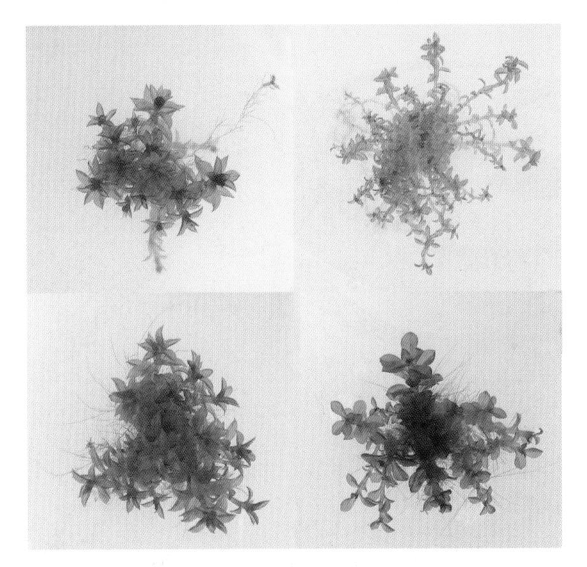

PLATE 40 Ecotypes of *Physcomitrella patens* showing the variations in form that appear within a species as responses to environmental conditions.

PLATE 41 English oak seedling (*Quercus robur*), root with mycorrhiza.

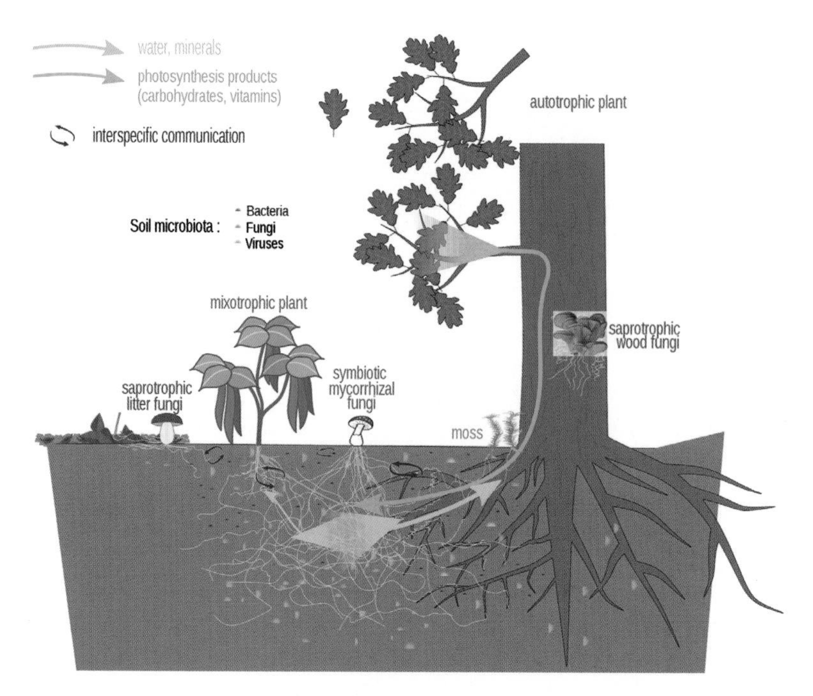

PLATE 42 Nutrient exchanges and communication between a mycorrhizal fungus and plants. Mycorrhizal fungi are in a symbiotic relationship with plants. The relationship is usually mutualistic, the fungus providing the plant with water and minerals from the soil and the plants providing the fungus with photosynthesis products. Some fungi are parasitic however, taking products from the plant without providing benefits. Conversely, some mixotrophic or parasitic plants connect with mycorrhizal fungi as a way to obtain photosynthesis products from other plants. Finally, saprotrophic (or saprophytic) fungi live on dead organic matter without establishing a symbiosis with plants.

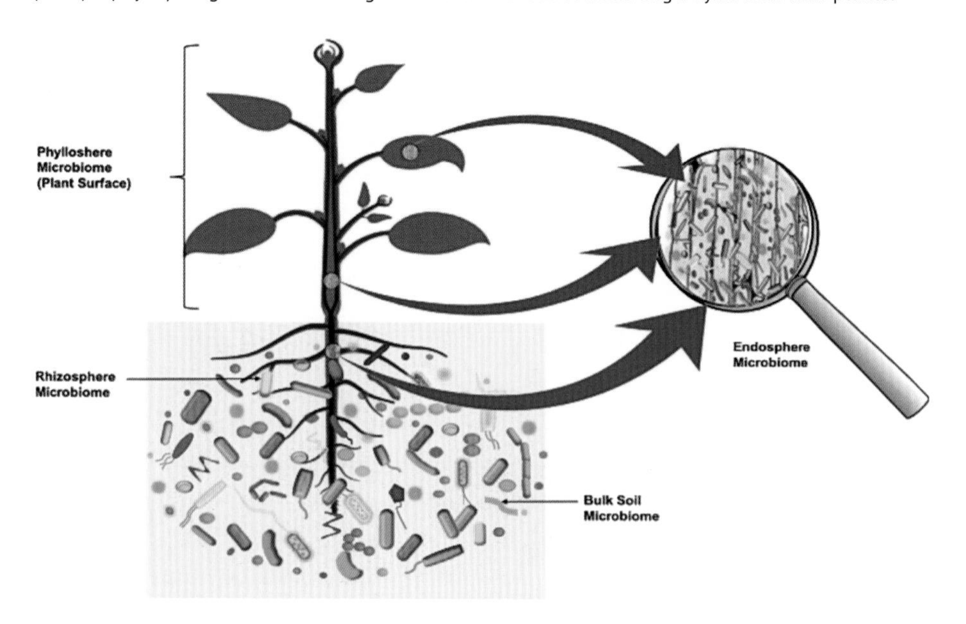

PLATE 43 The plant microbiome consists of diverse microbial communities on the outside surface and in internal tissues of the host plant. The rhizosphere, endosphere, and phyllosphere constitute the major compartments of the plant microbiome. The soil microbiome is an important source of the plant microbiome.

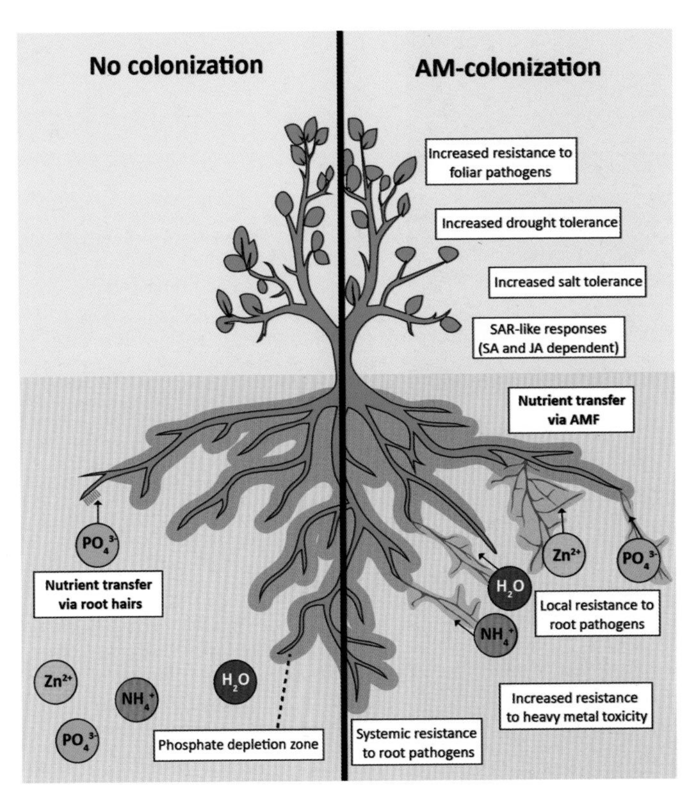

PLATE 44 Positive effects of arbuscular mycorrhizal (AM) colonization. The hyphal network of arbuscular mycorrhizal fungi (AMF) extends beyond the depletion zone (grey), accessing a greater area of soil for phosphate uptake. A mycorrhizal-phosphate depletion zone will also eventually form around AM hyphae (purple). Other nutrients that have enhanced assimilation in AM-roots include nitrogen (ammonium) and zinc. Benefits from colonization include tolerances to many abiotic and biotic stresses through induction of systemic acquired resistance (SAR).

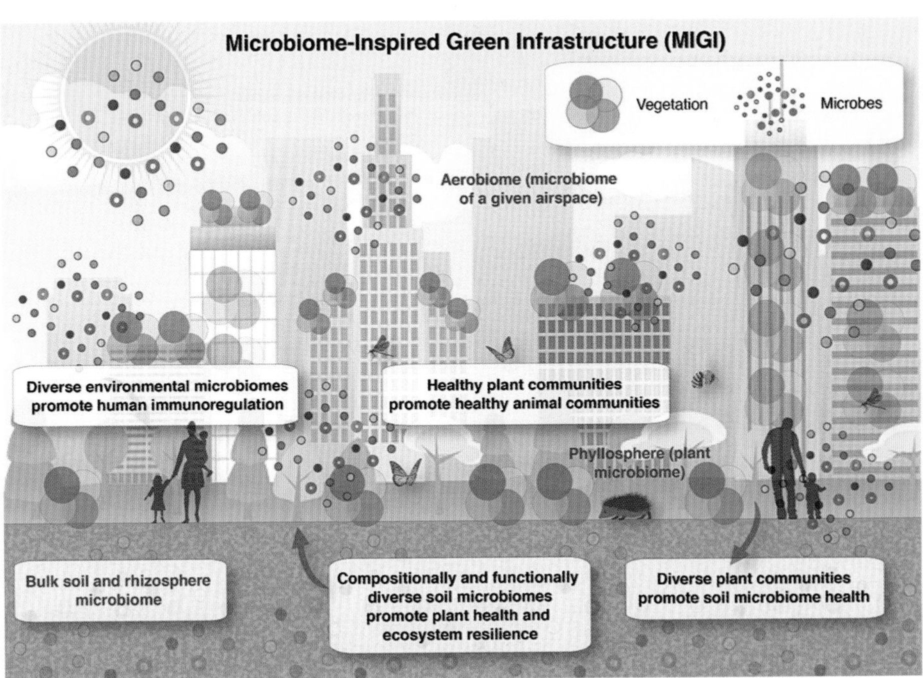

PLATE 45 Microbiome-inspired green infrastructure – urban multispecies health. Environmental microbiomes are the foundations of our ecosystems, and are essential to plant and animal health (including humans).

PLATE 46 Beyond plant community interactions and the immediate environmental conditions that affect plant behaviour, biocenology combines plant science with zoology to investigate the complex interactions of plants with other species.

PLATE 47 In Aotearoa (New Zealand), the Te Urewera forest changed from a national park to its own legal entity in 2014. It will own itself in perpetuity, with a board to speak as its voice and to provide governance and management.

PLATE 48 Mt Taranaki in Aotearoa (New Zealand) was accorded legal personality in 2017.

PLATE 49 (below) Manoomin wild rice in Minnesota has rights to exist, flourish, regenerate and evolve in a natural environment free from industrial pollution, human-caused climate change impacts, patenting, and contamination by genetically engineered organisms.

Kircher's idea was to use a carefully selected mix of perennials arranged in a random manner in a matrix. The rationale for this was to avoid the knowledge barrier associated with Hansen's approach, which required planting experience when establishing schemes in order to ensure that compatible species were laid out in the appropriate companionable combinations. Kircher's method didn't require a planting plan, but instead involved plant lists with quantities of various species chosen according to their habitat requirements, form, competitiveness and reproductive strategies. These can be arranged in any way to avoid creating a consciously designed planting and to offer a more naturalistic look: 'The idea behind this strategy is to create a plant community in an ecologically sound, competitive balance, comprising the ideal type of vegetation for public green spaces. Ideally, species showing various striking aspects, forms, heights, and propagation strategies complement each other to form a self-regulating system. Within this dynamic model the survival of the entire planting under extensive maintenance is more important than survival of individual plants.'[7]

The randomized method had been used during some prairie restoration projects in the United States, but combining it with Hansen's habitat and sociability criteria for plant selection made it unique. Other factors at play when choosing plants for the mix included practical considerations like growth traits and life expectancy, as well as aesthetic ones taking into account colour and texture, combined into an overall theme, generally colour based. The plant selections ensure long flowering times and compatibility between the species as they form self-regulating communities over time. Gradual progression means that some of the early establishing species disappear as others come to prominence. Despite the predetermined plants, the random nature of the arrangements, when placed in individual sites with different competitive conditions, means that each time they will look different, yet still maintain the overall community form.

Initial trials were carried out by Kircher at Anhalt University of Applied Sciences in Bernburg, along with LWG, Hemannshof and two other institutions: Lehr- und Versuchsanstalt für Gartenbau/Fachhochschule in Erfurt and Zürcher Hochschule für Angewandte Wissenschaften Institut für Umwelt und Natürliche Ressourcen in Wädenswil, Switzerland; together they form Arbeitskreis Pflanzenverwendung (study group for plant use and planting design). They have each contributed to the development of thirty-four different themed planting mixes for a diverse range of situations, with each mix trialled and assessed for at least five years. The German Perennial Nursery Association (Bund Deutscher Staudengärtner) has supported the endeavour and provides a retail interface for selling the mixes as tried, tested and ready to plant.

The make-up of the mixes breaks the plants down into percentages, with plants determined by their functional and visual roles as well as their CSR strategies. The design layer consists of competitive and stress-tolerant species in two groups. The first are 'Companion' plants, making up 30 to 40 per cent of the mix, recurring and creating the visual character of the community. The others are 'Dominant' plants, consisting of 5 to 15 per cent of the mix, and delivering a structural framework, particularly supplying vertical accents. The function layer provides a strong backbone to the scheme, with 'ground cover' plants, up to 30 cm/12 inches in height, comprising approximately half the mix, creating an anchoring base layer to provide a sense of unity and to exclude unwanted plants from intruding, while 5 to 10 per cent of short-lived 'filler' plants with ruderal characteristics offer immediate coverage in the first few years before being superseded by the other plants. 'Scatter' plants, usually bulbs, corms and tubers, give early season interest before the summer growth is established.

The first mix, produced in 2001, was Silbersommer (Silver Summer), featuring drought-tolerant species with an overall silver appearance. It was specifically designed to be low maintenance for dry, sunny, open locations on calcareous soil and featured, among the selection of twenty-eight plants, achillea, phlomis, sedum, aster, knautia and geranium, with alliums, crocus and tulips among the bulb species. Silbersommer contains the largest number of species, with the other mixes containing on average twelve to twenty. The Hermannshof prairie mixes – Indian Summer, Prairie Morning and Prairie Summer – distil the more technical approach of the garden into a simplified and more accessible form of their signature style. Variants on the method include consciously arranging the core group of plants within the mix in a considered pattern, using more than one mix and creating transitions between them, adding larger perennials or shrubs on bigger sites to create distinct focal points and interruptions, and combining the mixes with or allowing spontaneous species to be incorporated into the scheme.

The formulaic nature of the mixes takes the pressure off the users in making plant selections, and the randomized layout, evenly distributed across the area, simplifies the planting process, while management is an annual mechanized cut-back and removal of biomass. The ease of use and lower costs are particularly important for encouraging an uptake in public-realm planting, where they have found applications in a number of different situations, including traffic islands and roundabouts. Extending the reach of the mixes, Bettina Jaugstetter has been involved in promoting their use in a campaign in Weinheim with packages of smaller quantities aimed at encouraging local residents to purchase them from specified local retailers – as a bullet-proof garden in a box – to improve their own domestic spaces (Plates 22 and 23).

SPROUTING SPONTANEOUSLY

While perennials took centre stage within normal research institutes, the unique cultural context of postwar Berlin gave rise to a street-based approach to planting. The devastation of the city by Allied bombing, and its subsequent neglect due to political tensions, left large tracts of disturbed environments of rubble scattered in a mosaic pattern across the urban landscape. These wasteland areas, known as Brachen, provided ideal opportunities for adventitious ruderal plants to colonize in acts of primary succession. The Brachen hosted a wide variety of plants, both native and introduced, rare and common, spontaneously assembled in distinctly novel communities. Botanist Hildemar Scholz began to investigate these communities at Lützowplatz in 1953, noting the different species and patterns of vegetation, and kick-starting a field of research into what became termed *plantae urbanae* (urban plants).

Developing as a form of urban ecology, the spontaneous vegetation was analysed in terms of the ways in which it interacted with human activities to shape these spaces. The plant communities showed signs of varied fertility, influenced by factors such as discarded garden compost and proximity to roads, which encouraged forbs, while in nutrient-poor areas grasses dominated. Successional processes over decades revealed clearly transitioning biotypes, with larger sites hosting a variety of areas at different stages of development, maximizing biodiversity. Many of the Brachen plant communities failed to fit in neatly with the earlier-defined categories of plant sociology, and gave birth to a series of different ones located in new places, including railways, sports fields, airfields, streets and allotments.

Socially the spaces presented an alternative to formality and an openness suggestive of shrugging off the shackles of the recent past, with their unusual community compositions standing as metaphors for a new world of possibilities, and they were warmly embraced by local residents. Spontaneous vegetation provided an antidote to design, although interactions between the two began to develop as the city progressively regenerated. An early example at Teufelsberg in Grunewald, on the outskirts of the city, evolved as an artificial hill 120 metres/395 feet high, formed by 25 million cubic metres/27 million cubic yards of rubble exported from the inner parts of the city between 1950 and 1972. The site was both planted and allowed to develop spontaneously (Plate 24).

Enthusiastic campaigning to save the Brachen, gradually disappearing due to construction and development, resulted in the creation of the Natur-Park Schöneberger Südgelände in 2000 on the site of the former Tempelhof Station switching yards. Since closing in 1952, the area had been colonized by spontaneous vegetation, including grassland and woodland communities. The park has retained the planting in two distinct sections – the Landschaftsschutzgebiet (landscape protection area) and the Naturschutzgebiet (nature reserve) – which each have specific management strategies. In the latter section, urban ecologists at the Technische Universität (TU) Berlin work with Grün Berlin, a municipal service provider charged with managing the city's green spaces and infrastructure, to maintain a consciously naturalistic landscape to benefit biodiversity (Plate 25). This entails halting succession in order to keep certain areas open and free from trees to allow grasses and herbaceous plants to thrive. It is achieved by manually removing unwanted self-seeded plants and by grazing sheep. Raised boardwalks provide pathways that allow visitors passage through it without creating further disturbance to the vegetation. In the Landschaftsschutzgebiet, the vegetation is allowed to develop without deliberate intervention, but without predetermined pathways it is subject to random disturbance by people walking through it.

The management strategies ensure successional diversity in order to maximize the range of plant communities. Professor Ingo Kowarik , an urban ecologist at TU Berlin, describes the nature park as extending the hierarchy of the three traditional types of nature: 'A completely artificial site has been reconquered by nature. That means nature is different here than in other places. There's historical or primary nature of the first kind; second nature: cultivated landscapes, fields, meadows; third nature: urban green spaces, gardens, and parks. And then there's a fourth kind: the urban wilderness that has grown, more or less tamed, in these Brachen. This was the first time we realized we were dealing with a new kind of nature.'[8]

Kowarik first developed the notion of 'nature of the fourth kind' in 1991 after ten years of documenting the vegetation behaviour in the abandoned Gleisdreieck railway yards. Two decades later, following in the footsteps of Südgelande, TU Berlin and Grün Berlin were involved in transforming the Brachen into a new 26-hectare/64-acre park. Park am Gleisdreieck incorporated elements of the existing spontaneous vegetation into a more obviously designed landscape conceived by landscape architects Atelier LOIDL to provide a practical structure for users and a contrasting framework for the wild plants (Plate 26). While the Brachen were part of urban ecology's interaction between plants and people, this additional anthropocentric intervention provided a dialectical twist on the original idea, as geographer Matthew Gandy has suggested: 'In place of the original Brachen, we now encounter an elaborate staging of urban nature. Spontaneous biotopes have become the latest adjunct to landscape design.'[9] In contrast, a proposal to develop the grounds of the former airport at Tempelhof, which closed in 2008, was overturned by referendum, and an intended design to redevelop it as a park has been put on hold; it has been kept as an informal space with spontaneous vegetation, retaining the original ethics and ecology of Berlin's Brachen.

The former amusement park Spreepark in the Treptow-Köpenick district closed in 2002 and was acquired by the City of Berlin in 2014. Working with Grün Berlin, the authorities have been in the process of realizing a design by landscape architects Latz+Partner. In a similar manner to their design for the landscape park repurposing the former industrial site at Duisburg-Nord in the Ruhr, they will be retaining some of the existing spontaneous vegetation within the overall design of the park.

Norbert Kühn, a colleague of Kowarik at TU Berlin, has adapted the notion of nature of the fourth kind as a framework for active methods of planting, and has been investigating techniques for using spontaneous vegetation in creative ways in urban situations. Fully aware of the paradox of using randomly occurring plants in a designed context, he acknowledges: 'To intervene in spontaneous vegetation may seem contradictory: "spontaneous" means that which occurs by chance, without conscious design intent. We are dealing here with design using spontaneously occurring species. The starting point of this idea is to use plants that can clearly build stable communities under the given conditions of a site and to try to transform the plant communities according to a design perspective.'[10]

His rationale for using spontaneous vegetation in this way is fourfold: it incurs no cost, a factor important when public realm budgets, particularly for maintenance, are stretched; it is authentic not in the sense of being native, but because every site is unique in its community composition, giving it a specific sense of identity; it is appropriate to the site as it established there in the existing conditions, particularly in terms of dealing with stress and disturbance; and it provides important ecological processes in urban areas, both those performed by plants alone and those which also involve human activities.

Kühn recognizes that spontaneous vegetation needs to provide a certain degree of stability and continuity, ideally at an intermediate stage of succession, which offers more visual interest than just ruderal ground cover. In order to achieve this, certain interventions are needed: either to maintain the current state, as with selective cutting back and mowing; or, alternatively, to prevent any unnecessary disturbance in order to let processes continue unabated. Decisions need to be made regarding whether the cut biomass is to be left *in situ* or removed, according to the desired degree of soil fertility and intended visual appeal: 'To a large extent, managing spontaneous vegetation is similar to the management of natural areas: generally it is not a question of protecting one specific species of plant, but is more an attempt to prevent, manage, or speed up succession.'[11]

Kühn is well aware that, by its very nature, spontaneous vegetation is not conventionally selected or arranged, and therefore may not create the same aesthetic impact and seasonal interest as the public are used to from traditional plantings. To this end he has experimented, augmenting spontaneous communities with ornamental perennial plants selected to bolster the overall flowering effect and colour potential, to elongate its seasonal progression, or to enhance it in relation to its surroundings. This entails the insertion of particular species suited to the environment that have similar competitive traits to the existing species, because 'the success of such measures depends on the ability of the introduced species to coexist with the existing species – in other words, whether they can compete under the given conditions.'[12]

Trialling this enhancing strategy has involved using native plants as well as species from North American prairie and Eurasian steppe communities. Establishing from seed is unrealistic, as the introduced plants are unable to gain a purchase among the established species, so planting has been used instead. Kühn's results showed a marked ability of the steppe species to survive over the prairie plants, due to differences in ecological amplitudes, growth rates and timings. Outside of the trials, Kühn has yet to deploy this method of enriching spontaneous plant communities in a large-scale project. But with or without the insertion of additional species, he believes that the strategies of plants in unplanned communities can provide indications of how they are adapting to a changing climate, and thus give useful indications of the future of vegetation in urban areas.

NEW DIRECTIONS IN BRITAIN

While ecology and plant science had an early foothold in Britain, their influence in the creation of ecologically orientated landscapes took some time to emerge. Inspired by the developments in the Netherlands, an ecological approach to design finally began in the 1970s. Initial projects – including Central Forest Park in Stoke on Trent, the William Curtis Ecological Park next to Tower Bridge in London, and the Oakwood development in Warrington – aimed to provide access to naturalistic areas planted with predominantly native species. These projects generated a productive crossover between the fields of urban ecology and landscape architecture, providing practical solutions to the lack of habitats for urban biodiversity and economical alternatives to the high labour cost of traditional formal landscapes.

Britain's first ecological urban park, William Curtis Ecological Park, was named after the local eighteenth-century botanist and plant observer responsible for the publication *Flora Londinensis*. The project was spearheaded in 1977 by prominent conservationist Max Nicholson with his multidisciplinary environmental practice, Land Use Consultants, which had been involved in designing and realizing an ecological approach to the Central Forest Park in Stoke on Trent that opened in 1973. The Queen's Silver Jubilee Committee provided £4,000 funding on the understanding that the construction of the park would be less expensive than a traditional municipal one. A five-year lease on a peppercorn rent was obtained from the Greater London Council (GLC), overseen by the Ecological Parks Trust, which was responsible for managing and monitoring the project.

The 0.8 hectare/2-acre site on the bank of the River Thames next to Tower Bridge was a former lorry park with heavily compacted soil. In order to create a low-nutrient substrate, they introduced 350 truckloads of fill from construction sites, with volunteer help from the Conservation Corps and the Bermondsey Boys and Girls Brigade. The strategic addition of various organic materials allowed the site to be divided into a series of twenty distinct habitats, including meadow, a 480 sq m/574 sq yd pond, willow carr, grassland, woodland and scrub. The design was created by LUC employee Lyndis Cole, who co-wrote with Caroline Keen in 1976 an article in *Landscape Design* entitled 'Dutch techniques for the establishment of natural communities in urban areas'. The report on naturalistic planting projects in the Netherlands was instrumental in publicizing the ideas of their continental counterparts to a British audience, and clearly had an influence on the conception of the park.

Initial plantings included native species featuring 1,000 tree whips and were sown with grass seed, which were subsequently joined by numbers of ruderal species introduced with the seedbank in the imported landform media or by airborne means. The colonization of over 300 native and naturalized plants successfully provided habitats for a wide range of inhabitants, including 200 species of invertebrates, 28 species of birds and 21 species of butterflies. David Goode, senior ecologist at the GLC, monitored the project.

The park's location in such a dense urban environment proved to be extremely successful, attracting 13,000 visitors per year, and provided an important educational asset in constant demand for school visits. Its success fed into the ecological learning process of maintaining naturalistic sites in cities. Dealing with disturbance, compaction and water nutrient levels necessitated ongoing stewardship and study. Despite a three-year lease extension, the park closed in 1985 due to pressure from developers, but the lessons learnt from the experience fed into a number of other London projects, such as the Stave Hill Ecological Park and Lavender Pond Nature Park, both located in the former Surrey Commercial Dock in Rotherhithe; Tump 53 Nature Park in Thamesmead; and Camley Street Natural Park near King's Cross Station. The legacy of the park's ethos was also seen in the GLC's establishment of ecological principles in its planning and land management strategy developed by Goode.

NORTHERN LIGHTS

Another early adopter of Dutch ideas in the 1970s was Allan Ruff from the Department of Planning at University of Manchester. Ruff's enthusiasm was sparked by a field trip to the Netherlands in 1972, where he was particularly impressed by the ecological approach at the Gilles Estate in Delft, which stood out in stark contrast to the more traditional horticultural landscapes he was being shown. As he remarked, 'The Haydnlaan on the Gilles Estate was the most extraordinary innovative landscape, not simply in its use of native plants, but more importantly in providing a philosophical framework for a radically new approach to urban landscape.'[1] Revisiting to research further the following year, he was accompanied by Manchester graduate landscape architect Robert Tregay. They explored native plants at Amstelveen heempark, urban forestry planting at Amsterdam Bos Park and the Bijlmermeer housing suburb, and Louis Le Roy's endeavours at Heerenveen.

Tregay put these ideas into practice at the Oakwood Estate, a part of the Birchfield district in the Warrington New Town development in 1975. Hugh Cannings, Warrington New Town Development Corporation's chief architect and planner, had studied under McHarg in the USA, and was supportive of David Scott, chief landscape architect, and his deputy Tregay, in their ambitions 'to create a landscape in which ecological considerations . . . played a major role in design.'[2] The naturally functioning landscape was intended to provide the backdrop to the experiences of the residents' everyday lives.

The project involved the construction of a new housing community on the site of a former Royal Ordnance Factory, ROF Risley. Following demolition and clearance, the site was severely disturbed, with compacted soil, clay, rubble, shale and some contaminated topsoil. The decision was made not to introduce new soil, but to work with the existing conditions. Tregay outlined the strategy: 'You know that if you leave a rock quarry, a solid rock quarry, it'll be forest within a generation. You know if you just leave clay and rubble, it will become a forest. So I always knew that as long as it's not contaminated, a forest would grow. So why do you need top soil? You also know that low nutrient environments create more interesting ecologies than ones with top soil . . . So I just really applied that . . . planted it all as whips . . . and then, we got a forest out of it.'[3]

A survey had identified ruderal plants that had grown over the site during the thirty years it was derelict, although few were left after demolition. Those that remained, such as *Crataegus monogyna* and *Rubus fruticosus*, were incorporated into the design, and the survey was used as a guide for other species to introduce. Fragments of vegetation on the perimeter, including birch and willow, were stitched together with new trees to form a band of interconnected wooded belts; the idea was that they could 'be seen by people as a continuous thread of nature through their residential environment'.[4] Planting using a native mix of woody whips, including tree canopy species and shrub understorey, was distributed to achieve specific desired community effects in terms of diversity and complexity. Local nurseries had insufficient stock of the native species required, so the majority were eventually sourced from the Netherlands. This process was initiated three years before any building was undertaken on site in order to encourage healthy growth and natural form, and to create an established framework before residents arrived, as well as to keep costs low. A herbaceous layer was then planted and managed to reduce any unwanted growth, and coppicing and thinning of the trees maintained a balance between open and closed canopy structures.

The relationship to the housing estate was central to the project, balancing the ecology aspects with anthropocentric concerns, and framing it within the potential for development on brownfield sites. As Tregay explained in his 1984 paper, 'In search of greener towns', 'Behind the techniques lies a philosophy which is essentially humanistic. Natural design principles and ecological techniques are a means of achieving a richer, more diverse and more stimulating environment for people.'[5] This stimulation was also intended to act as a spur to encourage community involvement with the environment; the establishment of voluntary park rangers and children 'ranger helpers' not only kept a check on the parkland but assisted in its maintenance, including ongoing tree planting by local schools. It was also used for a wide variety of recreational activities. This form of engagement meant that there was less need for local authority manual workers, but rather employees adopted a more area-specific form of overview supervision.

Ruff and Tregay were responsible for disseminating these ecological approaches to a wider professional audience through a number of important publications, including Ruff's self-published *Holland and the Ecological Landscape* in 1979, and its expanded version in 1987, as well as *An Ecological Approach to Urban Landscape Design*, edited by Ruff and Tregay in 1982. Tregay and Swedish landscape architect Roland Gustavsson's *Oakwood's New Landscape* (1983) provided an update on the project's practical developments.

Due to the large scale of the early projects in Britain, their focus on the ecological processes of landscape and native plants meant that the finer detail of the science of strategies and behaviour of plant communities was not necessarily to the fore. However, the 24th Symposium of the British Ecological Society in 1983 saw a merging of the disciplines of landscape architecture, urban ecology and plant science, with papers by Tregay, Goode and Grime rounding out a bigger picture and suggesting new avenues to explore beyond a purely restoration-based approach.

THE SHEFFIELD SCHOOL

The work of Grime and plant biologist David Read at the University of Sheffield placed it at the forefront of ecological plant science, while O. L. Gilbert's investigations into vegetation dynamics and biodiversity in *The Ecology of Urban Habitats* (1989) explored cities as valuable habitats for a wide range of species. Integrating all of this into further research and practical applications, the Department of Landscape Architecture oversaw the development of a distinctly new approach to landscape design. Since the mid-1980s, its pioneering wave of research has led the charge in investigating the interface between ecological planting methods and the aesthetic enhancement of public spaces through the use of flowering plants. The two key figures of what has come to be known as the 'Sheffield School' of planting design are Professors James Hitchmough and Nigel Dunnett, whose work, both individually and jointly, has pushed the boundaries of planting in the public realm. Occupying the intersection of the disciplines of horticulture, ecology, design and psychology, they have ploughed a unique furrow in public planting, challenging antiquated ideas of traditional amenity thinking, and setting the ecological functions of plant communities within distinct aesthetic contexts.

Their work has addressed the dilemma of how to achieve the 'holy trinity' of landscape architecture by creating planting schemes that are aesthetically pleasing and tolerant of urban stresses, and function as ecosystems, when faced with the constraints and challenges of the twenty-first-century multicultural city. Constrictions in finances and skilled labour – not dissimilar from those that Robinson highlighted in the Victorian era – have made it difficult for local authorities in the UK to be able to ensure the upkeep of municipal parks and spaces. The ongoing ideological assault on local councils by consecutive Conservative governments since the 1980s has undermined their ability to provide quality green spaces for people just when their importance for health and well-being has finally been recognized. These restrictions, along with the need to address growing environmental concerns, create an imperative need for planting schemes that aim to be low-input yet deliver high-impact results, something that Hitchmough and Dunnett have been doing by creating new vocabularies of urban planting design.

To do this they have moved beyond using traditional conservation or restoration approaches, which they believe are not necessarily the most suited to city or suburban situations when creating new naturalistic style plantings. While recognizing the suitability of using native-only species in certain circumstances, they have embraced the fact that the built environment lacks any sense of specific natural vegetation communities, thereby creating opportunities for a more cosmopolitan approach to plant selection. The key criteria for selection, they asserted in their ground-breaking 2004 book, *The Dynamic Landscape*, are that 'plantings should have relatively low maintenance costs, be as sustainable as possible, taxonomically diverse, demonstrate marked seasonal change, and support as much wildlife as possible.' In order to achieve this it is necessary to 'move away from wholesale reliance on traditional horticulturally based plantings'.[6]

Key to this selection is an analytic approach to plants and their habitats, taking inspiration from a global selection of wild plant communities observed on numerous field trips combined with knowledge garnered through academic research: 'The central thrust in this work has been to take the understanding derived from contemporary ecological science and restoration ecology and apply this to the design of plant communities, and in particular, those composed of herbaceous plants (grasses, forbs, and geophytes) for use in urban landscapes.'[7] A key building block of the approach is the CSR theory of the processes that control the structure and composition of plant communities, developed by Philip Grime at the university and published in *Plant Strategies and Vegetation Processes*.

Despite the grounding of their work in the ecological science of plants, both Hitchmough and Dunnett place humans squarely at the centre of what they do. While plant communities must be robust and function well in terms of resistance and resilience, they also have to work hard to please people. Visual appeal is essential for the success of their projects, and the often-overwhelming use of abundantly colourful flowering is something of a trademark. There is a driving force to deliver the immediate well-being hits of dopamine that such planting offers, underwritten by the longer-term health benefits of making cities more liveable.

Their planting schemes are characterized by a strongly naturalistic appearance featuring seemingly random distribution patterns, and achieved largely through plants sown from seed *in situ*. The dynamic nature of this method allows patterns to emerge as plants grow, self-seed and find their own balances rather than through deliberate placement. With this in mind, careful species selection is vital in order to ensure that there is a sense of readability which the public can understand within what could be regarded as incongruously wild-looking landscapes in the formality of urban areas. The use of both native and introduced species plays an important role, both ecologically – to extend periods of flowering interest – and culturally for people, who may recognize some of the plants as species that are used in gardens, thereby giving the impression that this is in fact an intentionally planted area:

> Our experience to date suggests that in order to make naturalistic (irrespective of where the species come from) urban herbaceous vegetation acceptable to laypeople, it must be designed and managed to be visually dramatic at some point in its annual growth cycle. This often means increasing forb density and reducing grass density, and within the forbs increasing the density of the most dramatic, long-flowering species. If this means a designed native plant community looks substantially different from the wild occurring stereotype, so be it.[8]

Hitchmough and Dunnett's most high-profile project together has been the Queen Elizabeth Olympic Park in London. Working within the overall landscape framework for the park, designed by landscape architect George Hargreaves, and in collaboration with garden designer Sarah Price, they created floriferous, show-stopping displays of globally themed plant communities for the games held there in 2012. After the games they developed transitionary meadow-planting schemes to evolve into the current long-term park layout (Plate 27).

Despite their close association, they have each pioneered quite distinct takes on developing the relationship between ecological processes and dramatic aesthetics, which are justly recognized in their own right. What unites them – and sets them apart from some other contemporary practitioners of ecological planting – is a scientific rigour and an anti-dogmatic outlook.

While Hitchmough and Dunnett represent the most public face of the Sheffield School, the university has been a hotbed of research. The output from the Department of Landscape by Anna Jorgensen, Ross Cameron, Helen Hoyle, Nicola Dempsey, Harry Watkins and Jake Robinson has all helped to further investigate the various physical and psychological relationships between plants and people in the urban realm.

JAMES HITCHMOUGH

As his unique title of Professor of Horticultural Ecology suggests, James Hitchmough has carved out an academic niche that brings ecological science into a more understandable framework through high-profile projects in public places. Primarily anthropologically focused, his work aims to enrich human physical and psychological experiences while supporting as much other biodiversity as possible and minimizing collateral environmental damage. Working in the urban realm, his objective is to ensure that planting brings something extra to spaces beyond just amenity or aesthetic aspects. As he suggests, 'The work I do, it's certainly about ecology and I think it's about community too, it's about what people think about these things, and in particular I think it's about delight. So I'm interested in wherever I can, in creating experiences for people, even if they're temporal, which are really meaningful. So I'd like to see myself at least in having some role to play in what you may call the manufacture of joy.'[9]

But achieving joy is not simply a matter of producing crowd-pleasing fields of flowers, although much of Hitchmough's work does tick this box. Underlying it is an appreciation of the wider relationships of urban ecology, matching suitable species for specific sites with the cultural contexts in which they are situated. The criteria for these are based not simply upon location and environmental conditions, but also on local character, management strategies and social expectations, factored over the expected lifespan of the project. Anticipating these variables means matching situations that may necessitate a restricted maintenance regime with a context where a less manicured appearance would be considered culturally acceptable, while other places may demand a more controlled and neater appearance. These aspects play into plant selection based on species character traits, effective density required within the area, and the overall visual form of the community. Putting all of this into a framework to deliver cost- and resource-effective solutions, for both establishing and maintaining visually dramatic planting, is essential in order to ensure their success. Nothing sits more badly with clients and end users than a failed planting scheme.

As an ardent advocate of using a global assemblage of plants, Hitchmough describes his work as 'an eclectic mixture of species drawn from parallel habitats around the world, resorted to create designed cultural plant communities that flower and look dramatic for much longer than most naturally occurring plant communities . . . The vegetation is, however, always naturalistic, in that it has the visual patterns and rhythms found in semi-natural vegetation. It is also party to the same ecological processes that are inherent in semi-natural vegetation, as these processes are blind to the species present and their origins.'[10]

Working generally within a biogeographical mode, he has specialized in various natural plant communities with a specific focus on perennial-based meadows. As a typology, Hitchmough maintains it is ideally suited to delivering the attention-grabbing impact, and responses of delight, he aspires to provoke. Early work in the 1990s looked towards a Robinsonian effect by sowing native wildflower meadows into which were planted native and introduced species and looking at the competitive strategies of each. Since then his research and work have focused on plant communities from North American prairie and montane steppe, Western European meadow and woodland understorey, South African Mediterranean montane steppe and grassland, Eurasian steppe and Chinese meadows. Translating these into the temperate marine climate of Britain has not been a simple exercise of using imported species to create facsimiles, but has required a lot of research and analysis assessing each species' tolerance to local conditions. Some of these, such as the different moisture content between the native and the new locations, can mean some species are subject to mollusc predation, for which they have little evolved resistance, so looking for suitable proximate substitutions becomes a necessary part of the process. Compatibility of these new mixes is then subject to examination of their establishment rates and reactions to disturbance over five- to ten-year trial periods.

Putting into practice the knowledge gained from this research is a further part of the process. Prominent public projects, including at the Cambridge University Botanic Garden, the Oxford Botanic Garden, London's Horniman Museum and RHS Wisley in Surrey, showcase the results in botanically orientated arenas, appealing to those with an inclination to appreciate such detail, while the plantings at Burgess Park are arranged over a large-scale landscape canvas and subject to the wear and tear of daily life in South London. The plantings in the latter range from bioswales for drainage to prairie-style areas and large meadows on raised mounds, through which the well-worn desire lines of the public are much in evidence. These meadows feature a carefully woven and colourful mix of salvia, dianthus and origanum; their loose, wild appearance nonchalantly suggests an almost spontaneous community despite the large amount of introduced species familiar to garden settings (Plates 28–30).

His work has also been applied to more obviously designed commercial and domestic contexts in collaborations with landscape architect Tom Stuart-Smith, a designer more widely recognized for his contemporary garden style whose interest in ecological planting was piqued at University of Manchester, where he studied under Allan Ruff in the early 1980s. Their prairie collaborations at the corporate headquarters of Fidelity International in Tonbridge, Kent, and Stuart-Smith's own garden in Hertfordshire show how the approach can challenge traditional horticulture in terms of high visual impact.

Playing on the poetry of natural plant communities, Hitchmough's aim is to create changing events in space and time, with the drama of long-flowering periods that undergo visible weekly transformations. To achieve this, seasonal layers are employed from early spring ground cover to an early summer middle layer, followed by emergent later-season, tall forbs. The base layer blankets the ground, preventing invasion by unwanted plants. Bumps and rosettes form the middle layer, and clumps with tall, erect stems project above the foliage of the other layers, adding a transcendent drama as punctation and distinct visual markers. Maintaining a crucial balance between the numbers of species in each layer ensures suitable access for light while creating a level of ecological complexity. Getting this balance right is also crucial to ensuring resistance against unwanted invasion. Unlike natural meadows, which feature a higher ratio of grasses and are consequently more resistant to this, Hitchmough's mixes are composed primarily of perennials, so more susceptible but with the pay-off of greater visual impact. He is also investigating a more mixed savannah-like typology, with woody shrubs interspersed among the other perennial layers creating transparent visual interruptions on the level above.

Working on a public stage, persistence over time is essential for success, which means plantings have to be able to constantly resist weed invasion, something that densely layered communities help achieve, as well as featuring species of a broadly equivalent competitiveness. Consideration of the CSR ratio of species in the community composition plays a determining role in plant selection, according to Hitchmough: 'I nearly always work with species that are stress tolerators. Never work with invasive species, invasive species are no use to me because they would simply eliminate the diversity I'm trying to achieve, so I really want species that hold space but do not colonize space.'[11] Trialling species in experimental conditions has afforded him the knowledge of which ones will be most compatible together, avoiding a successional process that results in loss of diversity.

Another factor in choosing the make-up of the plant community is determining each species' 'niche breadth' – the theoretical range of environmental conditions that a species could successfully inhabit. Based on this information, the ideal spot in their ecological ranges is then ascertained where they can find their optimal performance when planted together while allowing for 'niche compression', which reduces resource availability when species are in close proximity. A sympathetic mix of species chosen on this basis can then be matched to the overall biomass capacity of the site, increasing its resistance to invasion, providing stability, and ultimately determining what the vegetation can become or do over time.

When it comes to methodology, Hitchmough uses sowing by seed not simply as a means of installation and establishment, but rather as a finely tuned design tool. His advocacy for seeding over planting is based on the advantage that it 'allows much more complex planting to be created with many species per unit area that flower for longer and create extremely fine-grained change on almost a daily basis'. In his book *Sowing Beauty* (2018), which maps out his system in detail, he asserts, 'Sown vegetation also brings improved ecological performance over planted, as the much higher density of plants makes the vegetation much more resistant to weed invasion.'[12]

His strategy involves three key considerations. First is establishing how many plants are required for the planting scheme, based upon what the intended outcome is visually, how long it will flower, and the ratios of specific flowering times of different species. Second is determining how many of each species are needed per square metre to achieve the desired result. Third is calculating how many seeds will be needed, based upon knowledge of the success rate of seeds to produce seedlings in amenable field conditions for germination, calculated per 100 seeds.

The seed is applied into a mulch of sand, gravel or crushed building rubble 50–75 mm/2–3 inches deep. After years of testing various different media, he has landed upon this as the substrate most reliable in ensuring slower but more even growth due to low nutrient levels, thereby reining in any undesired competitive tendencies of any of the species, as well as being highly resistant to unwanted ruderal invasion. It also ensures the seed bank in the subsoil is less likely to make an appearance detrimental to the sown species. Varying the composition of the mulch in mosaics across the site makes it possible to account for the detailed needs of different microclimates within a planting area, ensuring complete coverage with appropriate species in appropriate places while also increasing diversity.

Maintenance depends on the starting point – where initial expectations are determined at the design stage – but as these plant communities are dynamic, the directions they take are subject to various environmental vicissitudes and make management an ongoing responsive process without a set book of rules: 'You can't write a specification for it, you have to look at what is in front of you and you have to process what's going on, you have to assemble some goals for what you want it to be and you have to try to work toward that through various interventions. It's hard, it's intellectually quite challenging to do that, it requires quite a lot of processing power, because it requires the capacity to look into the future.'[13] It is the combination of this considered forethought with the ongoing real-time interactions of the planting which makes these projects exciting experiments in the art of sustainable urban ecology.

To this end, Hitchmough also devotes time to anticipating the effects of climate change in order to build resilience into his plant mixes, mapping predicted temperature changes to potential species selection for trialling. He sees this as creative rather than restrictive: 'Climate change also helps to free up conventional thinking by making people come to terms with the idea that the future will not be the same as the past.'[14] It opens up further possibilities for his vision of novel mixed-planting ecosystems composed from a global palette of plants that can cope with the new challenging conditions that are coming.

NIGEL DUNNETT

Nigel Dunnett is something of an urban evangelist with a driving passion to bring naturally inspired experiences to people in cities: 'It's not something that's an option, it's something that is a must have . . . Because we need to be bringing back soils and nature and vegetation, and the interactions between them, into the heart of our cities, into the heart of the places where people are living and using, to combat those challenges of climate change and urbanization. And bringing back nature into cities is the only way to do it, but the question is what type of nature.'[15]

His response to his own rhetorical question is something that he calls 'Future Nature'. His view is that previous ecologically orientated efforts based on a conservationist approach have failed to deliver something that is more than just functional in cities. This failure, he contends, rests simply in only delivering benefits on one level of urban ecology, to the neglect of more social and cultural ones, and lacks a language for people to understand them and their positive aspects. Instead, Dunnett's approach is inspired by natural patterns and processes, and is based on a visual ecology referencing well-defined plant typologies such as meadow, prairie, steppe, savannah and woodland. These provide starting points for designed communities in which novel combinations of native and introduced plants are arranged to produce flower-rich landscapes. Being influenced by natural plant communities, but not beholden to replicate them faithfully, has provided him with a certain freedom and an opening up of creative possibilities, to try to hit the sweet spot where the benefits of planting for biodiversity and people coincide. Accentuating the latter has been a key strategy for delivering the former, by using urban planting as a lever to generate positive psychological responses and uplifting experiences.

This is expressed in his work, in a manner akin to pop art, by using big, bright and bold planting as something to attract people and get them to engage with plants, even if it is only for an Instagram opportunity. But he believes that each encounter is meaningful and satisfying on deeper levels. Describing it as 'enhanced nature', he argues that it is 'an art form, not because of the aesthetic so much or the way it looks, but because of the way it makes us feel, and I think when we work with landscapes tuned to nature, it reaches deep inside us to that buried, inbuilt connection to the wild that we have, and which so many people have lost.'[16]

After early work on green roofs and rain gardens, he developed the idea of creating impressionistically colour-laden meadow schemes of annual flowers that import a heightened type of rural sensibility into the densely built environments of cities. As temporary low-input solutions to landscapes in transition, they delivered a high visual impact, which encouraged engagement from local residents. Spurred on by the enthusiastic responses, Dunnett developed a number of colour-themed seed mixes of perennial and annual plants, and in 2004 he helped to establish the company Pictorial Meadows with the University of Sheffield and social enterprise Green Estate, supplying local authorities and large estates as well as the general public. These perennial meadows deviate from semi-natural meadows in their exclusion of grasses, whose competitive tendencies lead to the gradual disappearance of the flowering forbs, thereby allowing a reliably ongoing plant community robust enough to withstand the disturbances encountered in urban areas.

His high-profile public projects include a large woodland at Trentham in Staffordshire and the Barbican in London. The latter, being a podium landscape built above a car park, necessitated shallow planting media in order to reduce weight loadings. The planting transitions from a shaded deciduous woodland area to an open steppe-style landscape consisting of grass and perennials suited to dry conditions and chosen to deliver seasonal waves of colour and visual interest in winter (Plate 31). The Grey to Green Project in Sheffield transformed some of the inner-city pavements into a sequence of interconnected bioswales, planted with species featuring a variety of ecological ranges in order to accommodate the moisture gradients from the drier top to the wetter bottom of the swales, and chosen to deliver visually in such a prominent location.

Dealing with a diversity of environments, Dunnett uses a general planting strategy based upon something he calls 'Universal Flow', which combines the ecological traits and tendencies of plants with a spatial and temporal framework. Plant selection relates to naturalistic community typologies but takes a mixed biogeographical approach, with consideration given to the compatibility of species based upon their reproductive capacities and growth rates. Their tendencies either to remain fixed in one space or to spread by seed or clonal reproduction, and the vigour and speed with which they carry out these functions, provide indications for their application and arrangement. Dunnett's PhD supervisor at Sheffield was Philip Grime, so it is not surprising that he uses the CSR theory when considering selection, taking into account the relationships between stress and disturbance along with factors such as anticipated productivity, stability and diversity within the plant community. Generally avoiding overly competitive species is key in preventing dominance, and ensuring a moderate degree of stress and/or disturbance is present allows dynamic forces to play out within a setting of relative stability. The balance of species within a perennial meadow scheme features a predominance of fairly static plants with moderate competitive traits interspersed with some weaker spreaders and seeders, while a temporary planting would be based on an inverse relationship.

Planting arrangement takes into consideration the forms of plants and the effects of naturalism they evoke when organized both vertically and horizontally, using an embellished form of matrix planting. Elements of order are subtly introduced into the random spatial arrangement through layering and patterns of distribution which reference semi-natural communities. Dunnett's key principle is one he describes in the book outlining his methodology, *Naturalistic Planting Design* (2019), as 'centres of gravity', in which agglomerations of a species that provide visual anchors are accentuated by outliers that thread themselves through other species. These create forms of flows and drifts and introduce a sense of repetition and rhythm across an area of planting. A decidedly aesthetic angle is introduced through this patterning, particularly in relation to seasonal change. He imposes restrictions to ensure that only three plants are flowering at any one time in order to simplify the complexity and ensure a sense of readability. The use of colour is central to this, because 'carefully chosen colour can lift the scheme to another level entirely.'[17]

In this respect, he departs from most other ecological practitioners in his keenness to highlight the self-awareness of the design process, and is not afraid to celebrate planting as an art form. He is also not shy of admitting certain historical influences. He acknowledges that the painterly aspect of colour in his work is an inheritance from Jekyll updated for the digital age, with the patterning he employs as a form of floral pixelation.

Management of his planting schemes requires intervention at points of increasing or decreasing diversity to prevent dominance of competitive species, or to allow increases through self-seeding, in order to maintain the desired community balance.

GLOBAL GARDENING IN FRANCE

GILLES CLÉMENT: THE GARDEN IN MOVEMENT

Since the 1980s, landscape architect Gilles Clément has been the most prominent proponent of ecologically orientated planting in France. Critical of the damage done by the anthropocentric world view, his thinking is anchored by a humanist ecology which situates humans within the natural world, as actors in ecosystems with the ability to rein in their interventions, to ensure a more equitable balance with the rest of nature: 'It is that which in some way is going to improve human life [and] allow us to live in harmony with our environment: a give and take. Humans are taken into account, integrated in the whole.'[1]

He sees the practice of gardening as much about what is done on a small plot as on the level of the whole planet, in which the global interconnectedness between people and plants is an ongoing process of co-evolution. Unlike conservationists or native plant protectionists, he rejects both physical and ideological nationalistic boundaries, instead embracing the new form of diversity that migration offers. He also challenges the purist value judgements that create hierarchies between different types of ecosystems, instead appreciating near pristine environments as much as derelict, urban, brownfield sites. Learning lessons from both, his work champions the creation of novel ecosystems in often unique contexts that aim to maximize biodiversity and deliver a balance of aesthetic and educational outcomes.

Clément's projects are based firmly on empirical evidence; he works with natural processes and never against them, preferring to be guided by plants rather than corralling them into aesthetic assemblages according to taste. His interest lies not in the characteristics of particular plants or the artistic possibilities offered by arranging plant communities, but instead in establishing the contexts in which plants can develop after establishing. Like most other practitioners, he has learnt much from looking at plants in natural habitats around the world, as well as paying close attention to his own projects as they develop. Celebrating observation over intervention, his work reduces gardening activities to the minimum, while still recognizing the creative possibilities that sensitive stewardship can offer. The intended outcome is less predetermined in visual appearance and form than traditional design, as his embrace of the dynamic and stochastic activities of natural processes looks towards plants' adaptive capacities, resilience and successional change to determine the end result. Consequently, there is never a sense of failure in a project, just an ongoing series of changes and interactions from which there is much to be learned and appreciated.

Active and outspoken on environmental issues, he has never been afraid of new ideas or communicating them to the public, which, along with many high-profile projects, has gained him a highly regarded position in his homeland. Yet despite being a frontrunner in ecological landscape design, Clément is something of an outlier in terms of international recognition, possibly as his experience is more practice-based than the research- and science-based approaches of the British and German practitioners. His methodology is based on close attention to projects after they are completed rather than before: 'I'm very pragmatic. I have my garden which has taught me practically everything, and based on that experience, I see how we can extrapolate, how we can use these elements in theorizing, and apply those theories to different scales, to various social contexts in different places. So basically, I devise a theory "after" experimenting. And the essence of what I've been able to discover, among other things, which leads to conceptual principles, from a methodological point of view . . . is to work as much as possible with nature and as little as possible against it.'[2]

This principle is embodied in the first of three key concepts that run throughout his work. The idea of 'the garden in movement' was initially proposed in an article published in 1984 under the title 'La Friche apprivoisée' (The tamed wasteland) and then elaborated in his book *Le Jardin en mouvement* (1991). It is concerned with working 'with the energy that nature offers us. It's there. It's extremely economical to use this energy, by directing it, bending it a little when it is absolutely essential for us, but without fighting with it.'[3]

The idea 'originates in the physical migration of vegetal species within a given area, which the gardener interprets at will', and proposes observing more and gardening less in order to understand the site-specific interactions between plants over time when left to develop without much interference. Design becomes a process not so much of dictating an outcome as stepping back and deferring to the nomadic tendencies of plants to self-seed, colonize and move around a site. Rather than attempt to predetermine the competitive tendencies of different species to achieve relative stability in a plant community, Clément prefers to let the plants do this themselves in real time. Succession is allowed to take place at its own pace, with changes in the composition of communities generally accepted but carefully monitored. Maintenance becomes a practical, hands-on, light-touch process instead of being predetermined at the drawing board. Intervention is intended to ensure maximum diversity by creatively channelling competition between plants and selective pruning. The role of the gardener is to work with the energies present in the garden to realize their potential, and to work around self-seeded plants, making decisions to leave as many as possible where they have established, in order to maximize biodiversity. Clément expanded upon this idea in *Éloge des vagabondes* (In Praise of Vagabonds, 2002).

This principle has been crucial in the development of Clément's personal ongoing masterwork: his garden La Vallée in Creuse. Purchasing some land in 1977, he built his own house before turning his attention to developing the 5-hectare/12-acre site, intent on maintaining and increasing the biological quality of the substrate and water, changing as little as possible and completely avoiding fertilizers and pesticides. Observing the processes of nature, ignoring the concept of weeds, and working responsively with the site as it evolves through time have been key to its success. He has shaped spaces by selectively removing plants in order to let others follow their due course and create new ecosystems; the garden changes annually and Clément has adapted accordingly, redirecting paths around plants that have sprouted up and leaving fallen trees to regrow from their new positions.

The idea of the garden in movement was applied to a large-scale project that put Clément in the public limelight and which, to date, remains his most acclaimed. Parc André Citroën, co-designed with landscape architect Alain Provost and architect Patrick Berger, opened in 1992. The design suggests a narrative journey from the naturalistic to the artificial; contrasting areas are left to create their own equilibrium with large areas of geometric formality, and a series of colour-themed gardens transition through the spectrum to black and white gardens at the rear. The Garden in Movement near the Seine riverfront is continually reshaped by various factors, including the gardeners, the public creating pathways through the space, and the plants themselves, giving it a highly dynamic atmosphere (Plate 33).

Other projects employing the strategy include the 3-hectare/7.5-acre steppe-style landscape of the Escalier Jardin at Foyer Laekenois in Brussels and the 5-hectare/12-acre parkland at the École Normale Supérieure in Lyon, which deploys sheep as part of an organic cycle maintaining the meadows and providing fertilizer for the plants. The 2-hectare/4.5-acre gardens surrounding architect Jean Nouvel's museum building at Quai Branly in Paris provided another opportunity to develop the idea, but with a savannah-style environment based upon a large-scale planting of miscanthus.

Thinking philosophically about landscape and ecology is key to Clément's work. His second concept, 'the planetary garden', extends the traditional idea of the garden to a global level to embrace the planet as a whole: 'Instead of being limited to a small space that we control, from now on the garden is placed within the limits of the biosphere.'[4] The genesis of this perspective was the publication of the original NASA photos of Earth from space. This epiphanic event led Clément to an understanding of the finite interconnectivity of life on Earth, and he drew from this a parallel between the planet's precariously contained biomass and the original idea of the garden: the *hortus conclusus* or *garten*, an enclosed space protecting things that are precious and valued from the vicissitudes of the wilderness. Recognizing the overwhelming impact of humans on the planet, he sounds a clarion call encouraging all of us to become gardeners tasked with increasing and maintaining biodiversity. He defines this task through nine actions: '1. Not harming the earth; 2. Welcoming the gardener's allies; 3. Promoting the exchange between living creatures; 4. Knowing how to manage water; 5. Building the human house; 6. Preserving the gardener's enclosure; 7. Caring for the earth; 8. Giving nature its share; 9. Producing without exhausting.'[5]

On a global level, he places prime importance not only on preserving the world's biodiversity by protecting hotspots, but also in creating conditions in which it can sustainably flourish and evolve by using plants from around the world in biogeographically compatible environments. Fully aware of the diversity benefits of endemic plant communities, he also acknowledges the basic ecological principles that can create new ones through the processes of plant migration by human, animal and weather flows: 'Intermingling creates the conditions for a new landscape and threatens diversity, increasing the specific richness on a local scale and reducing it globally. Adding and subtracting at the same time. The nature of what has been subtracted is known to us: the total number of creatures living on the plant at the moment is diminishing. The nature of what has been added is less clear.'[6] It is a proactive vision of the world and, for Clément, a political one; he rejects the prevalent ethos of the Anthropocene, which views the planet as a stock of resource commodities to be used without regard for consequences.

Clément's final concept, 'the third landscape', takes 'the garden in movement' one step further by looking at the dynamic tendencies of plants to shape environments unaided by human assistance. It is inspired by the colonizing characteristics of plants to establish themselves in unlikely and often inhospitable places, taking advantage of disturbance and low nutrient availability, such as in abandoned urban areas and neglected rural land. The colonization of these marginal zones highlights the potential ecological value for biodiversity of such sites, as well as the tenacious migratory tendencies and adaptive characteristics of ruderal plants, which make the most of short-term opportunities.

In his *Manifeste du tiers paysage* (2004), Clément situates 'the third landscape' between primary locations that have not been subject to human exploitation or major intervention, and as such are diverse and relatively stable ecosystems – such as forests, alpine meadows, moors and tundras – and those that have been consciously constructed or heavily managed, such as agricultural land which is monocultural and species poor. The third landscape 'lies somewhere else, in the places that are neglected: the sides of the road, abandoned plots, *friches*, heath, and peat bogs – anywhere where it is difficult to exploit the land with machines. It is the sum total of these that I call the "Third Landscape", a precious assembly, if you consider what it represents as a legacy of genetic diversity.'[7] These areas may be ecotones, transitioning between two biological communities, or 'interfaces, canopies, edges, borders' where 'their richness is often greater than that of the environments they separate.'[8]

Parc Matisse in Lille neatly embodies the thinking behind the idea. The park, designed in collaboration with Éric Berlin, Claude Courtecuisse and Sylvain Flipo, comprises an expansive open grass area, not unlike a traditional park, featuring an impressive 0.35-hectare/1-acre 'island' raised 7 metres/23 feet above ground level and inaccessible to the public. While not actually an abandoned site, the earthwork was created with the local authorities as a space to monitor the progress of pioneer species in an 'ideal forest'. The feature was named Île Derborence in a nod of recognition to Derborence Forest in Switzerland, which is one of the few European primary forests largely unaltered by human intervention. Its form was inspired by the uninhabited Antipodes Islands of Aoterea (New Zealand). The landform was created from a former hill that was cut and remodelled using material excavated during the construction of the Eurostar/TGV Lille Europe station, and contained within rough cliff-like concrete walls (Plate 32).

The flat plateau on top was planted with a mixture of European boreal biome species and Asian and American plants. The plants have been left to create their own balanced ecosystem, and are visited only twice a year by the head gardener to analyse its development. Initial responses from the neighbouring residents to such a dramatic imposition were negative, and the island had to have trees planted around it at ground level to soften the brutalist aspects of its facades. But through gradual education about the aims and processes of the project, the neighbours have been won around to it. Clément has continued working with the local authorities to monitor the developments, utilizing the results in the management of the rest of the park.

The concept also informed Clément's design at La Défense in Paris, in which an abandoned stretch of motorway was repurposed to create a horticulturally rich linear park in the dense urban fabric. The shady aspect of the overpass and the polluted atmosphere favoured an unusual approach to planting, and he rejected conventional ideas of city plants in favour of those compatible with the site conditions. Gunneras play a disruptive architectural role along a promenade stretch, while buxus, gaura, pennisetum, corylopsis and hamamelis all play starring roles in the scheme.

Perhaps the most comprehensive project bringing together Clément's concepts is one developed between 2009 and 2012 as part of the Estuaire Nantes/Saint-Nazaire art festival at a unique location in a former military base at the port of Saint-Nazaire. German-built and used as a dock for submarines, the large concrete structure was designed to be bomb-proof, although by the end of the Second World War it was in three incomplete sections. Addressing each area separately, Clément developed three distinct ecologies: Bois des Trembles, Jardin des Orpins et des Graminées and Jardin des Étiquettes (Plate 34).

The first, on the corrugated roof of the dock, was planted with an arbour of aspen, *Populus tremula*, spaced in an almost grid-like manner across the surface; the trees' roots were tethered, with their upper forms free to move and tremble in response to the wind across the exposed site. The second intervention features a 'garden in movement' of self-seeding plants, including *Stipa tenuissima*, *Euphorbia characias* subsp. *wulfenii* and *Centranthus ruber*, which have been left to colourfully colonize a lower level of the building's roof. The third area required the importation of a thin surface layer of soil and gravel in which plants have appeared from seeds delivered by the wind, birds and passing shoes. Plants are allowed to set seed and grow, and twice a year the newcomers are identified and labelled, creating a dynamically changing botanical garden.

OSSART AND MAURIÈRES: A WATERWISE PERSPECTIVE

Key to planting ecologically is not only observing overarching principles of plant communities but also paying close attention to local specificities. The responses of plants to geological and climatic conditions, and their co-evolutionary dependencies on humans and other species within specific environments, require different regional responses that acknowledge the unique conditions and characteristics of the terrain.

The major terrestrial biomes found across the planet accommodate a range of different habitats for plants. Due to heat and aridity, the Mediterranean biome is a particularly challenging environment for plants, and over time species have adapted in unique ways that define their life cycles and forms. Elemental forces, such as periodic fires and moisture availability, have played a role in composing the spatial arrangement of plant communities in Mediterranean landscapes, and intense direct and reflected solar radiation have produced defensive characteristics in many species. Some have developed glaucous foliage, white hairs and light-diffusing surfaces, while other species release aromatic oils to reduce transpiration and deter predation. Unlike in temperate regions, most growth takes place in autumn and often through the winter, with a short flowering season in spring followed by summer dormancy.

The Mediterranean basin hosts a potential plant palette of 25,000 species, one-tenth of the world's flora, yet only a few hundred plants are frequently used in gardens within the biome. For many gardeners, the Edenic ideal of the garden suggests a lush abundance of soft greenery, which is antithetical to the rougher subshrub and succulent species that populate the vegetation of Mediterranean environments. The original paradise gardens were a retreat from their arid surroundings into verdant spaces, and this attitude still lingers in garden ideals today. Unfortunately, many gardeners in dry climates work against the constraints of such conditions, relying instead on irrigation to compensate for lack of moisture, often in order to support plants from wetter parts of the world. Not only is this reliance on water increasingly unsustainable, but it also reflects an anthropocentric arrogance in the face of environmental limits. It is common across the wider Mediterranean biome, from large green swathes of lawn in Los Angeles to roadside rose beds in Dubai. Planting in arid areas is all too often a display of the ability to dominate and manipulate the environmental conditions, broadcasting the fact that the means and wealth are available to do so. The over-reliance on irrigation creates a vicious cycle in which native plants, when they are used in gardens, are overwatered as a matter of habit and don't survive: they are susceptible to disease from too much moisture, thus tarnishing them with an unreliable reputation. As a consequence, they are less popular, so nurseries are reluctant to stock them due to the economic necessities of supply and demand. Imported plants, which require constant water to survive, are then used more in a self-fulfilling and unsustainable loop.

Design duo Eric Ossart and Arnaud Maurières have used plants appropriate for arid areas in the gardens they have created in Mediterranean biome contexts across the continents. Their book *Éloge de l'aridité: Un autre jardin est possible* (In praise of aridity: Another garden is possible, 2016) sets out a manifesto for changing the paradigm of gardening, framed as it has been for the past two hundred years around the hegemony of British gardens: 'The English climate is not that of the whole planet: there are many other gardens to imagine in regions where it rains much less. A garden without watering is no less a garden: it's time to adapt the aesthetics of the garden to respect the climate of each place – however arid it may be – and to no longer confuse exuberance with an abundance of water (an agave steppe is as generous as a cluster of roses).'[9]

Carefully considered plant selection is key to their aim 'to abandon resolutely irrigation as the *sine qua non* for the garden to exist'.[10] Understanding the morphological adaptations and strategies that allow plants to survive for long periods without water provides insights into how they can best be used within a garden context that respects their ecological tolerances. Also understanding that aridity is not simply a lack of water – it is also a lack of water at certain times of the year – is important in relation to growth and flowering cycles. Seasonal patterns of precipitation, temperature ranges over the year, and even between day and night, play into the particular determinants for plant palettes for specific regions and local microclimates.

The work of Ossart and Maurières has taken both a biogeographical and a nativist approach to planting. On various projects in Morocco, where the native vegetation is limited, they utilized a wide global range of Mediterranean plants, many collected as seeds or cuttings from places such as Madagascar and Yemen, to create landscapes of seemingly spontaneous appearance, including naturalistic meadows composed of species of American agaves, African euphorbias and Saharan grasses. However, at Los Garambullos, a project near San Miguel de Allende on the Mexican altiplano, sensitivity to the endemic flora was the determining factor of the design strategy: 'In Mexico, we have all plants we need on site, and our intervention is much less artificial. We deal with the existing flora on site, which is already very rich. And we are expanding to other wild plants that we harvest the seeds from in the neighbouring regions. We give the seeds to local nurseries that provide us with young plants. We are planting in large quantities to give a very spontaneous appearance to the new introductions.'[11]

The rich indigenous flora, composed of shrubs such as mesquite (*Prosopis* spp.) and cacti (*Opuntia* spp. and *Myrtillocactus geometrizans*), is interspersed within a natural matrix of *Rhynchelytrum roseum* grasses, and accented with other cacti, agaves, salvias and shrubs, ensuring a density to the community so that it is able to resist invasion. Unlike restoration ecology, which neglects a designed aspect, Ossart and Maurières employ a creative angle in community arrangement and take their lead directly from the existing vegetation in order to accentuate it. Creating layers is essential in order to enhance symbiotic associations between plants in the community, particularly in respect of water requirements. The planted species do not need to compete with their surrounds, as they complement what is already there. While a certain sense of intervention is apparent, subtlety is essential in order to ensure that the boundaries of the property blend unnoticeably with the surrounding landscape (Plate 35).

OLIVIER FILIPPI: BRINGING THE MEDITERRANEAN HOME

Working against conservative attitudes of gardeners in the South of France, French nurseryman Olivier Filippi has been attempting to disrupt their complacency and encourage them to open up their imaginations to the possibilities offered by the wide yet unexplored range of endemic plants. His mission is the adoption of a new mindset by gardeners that appreciates the importance and beauty of native plants, and recognizes they are the best suited to and most sustainable for the area, as well as crucial for reducing resource and maintenance inputs.

Learning from the landscape is an essential part of Filippi's process of understanding local plants and appreciating the ways in which communities arrange themselves. His foray into the flora of the Mediterranean basin has involved decades of studying, growing and selling plants with his wife, Clara. Trips around the basin, from Croatia to Crete, Morocco to Greece and Sicily to Spain, have yielded a considerable body of knowledge relating to the ecological processes of the landscape. Fundamental to Filippi's approach is overturning many of the traditional horticultural assumptions about gardens, such as improving soil and mulching, as well as re-evaluating typical expectations of seasonal patterns and plant flowering, and calling into question the usual role of the gardener. In particular, he discourages the use of irrigation, and is insistent that selecting ecologically appropriate plants is a tried and tested method for sustainable gardens: 'Watering indiscriminately eliminates plants from dry land. I try to make people understand that we can refer to the four-thousand-year history of gardening around the Mediterranean, while the use of drip, lawn sprinklers, and automatic watering appeared barely fifty years ago.'[12]

In his own garden in Mèze, he has spent many years experimenting with ways to work symbiotically with regional plants, relating the process to techniques used in agroforestry management. Shaped over millennia through clearance and farming, the Mediterranean landscape contains very few remaining wild plant communities. Human intervention has been a major driver in shaping the vegetation of the area, particularly through managed animal grazing, which creates disturbance and reduces the possibility of certain species dominating. The animals also aid in the distribution of plants by disseminating seeds across distances. For Filippi, such human influence on the landscape is a sign of symbiotic engagement which can be translated into the management practices of gardens.

Inspired by the characteristic open spacing and mostly evergreen low-height vegetation of the fragmentary garrigue scrublands found on calcareous plateaux in the region, Filippi has developed planting methods based upon plant adaptations and responses to the challenging panoply of conditions they face. Prolonged heat and drought, bouts of cold and dry winds, salt, fire and herbivore disturbances have all played roles in developing unique characteristics. An understanding of the co-evolutionary strategies developed by plants to deal with these factors has afforded Filippi insights into the diversity and ingenuity of plants in such inhospitable surroundings and the ways in which they interact with each other.

The cyclical recurrence of fire has been an integral part of Mediterranean ecosystems since at least the Neolithic period, and the consequent disturbances caused by regular outbreaks are part of the auto-successional nature of the vegetation. Fire impacts on plant communities in different ways. Some species are destroyed or weakened by it, some are unperturbed by it, while others actually benefit from it, and are evolved to use its effects to their advantage. Those resistant to fire are known as passive pyrophytes, while those that use it in order to assist in spreading their seeds are active pyrophytes. The latter benefit from the devastation caused by fires, as the reduction of foliage within the community after a burning eliminates competition and provides more surface area on which their seeds can fall and generate. The Montpelier rock rose, *Cistus monspeliensis*, is one species whose seeds are ruptured by fire, while the smoke and heat trigger germination. Other plants, such as arbutus and phillyrea, have lignotubers: large organs of woody, rounded outgrowths surrounding the base of the trunk which contain a mass of vegetative buds, vascular tissue and substantial food reserves. If the upper part of the plant is destroyed, it is renewed by rigorous regrowth from the lignotuber.

Another local adaptation is to be found in the antisocial antics of allelopathic plants, which excrete biochemical substances to slow or stop the growth of their neighbours and prevent seeds from germinating in order to reduce competition for scarce resources and ensure their own survival. The term allelopathy, derived from the Greek words 'allelo' (one another or mutual) and 'pathy' (suffering), was coined by Austrian professor Hans Molisch in his 1937 book *The Effect of Plants on Each Other*. These chemicals are given off by different parts of the plant – excreted from the root or emitted as airborne volatile organic compounds – or they can be released through natural decomposition. Allelopathy is an important determinant in the distribution and make-up of plant communities, characteristically giving them the appearance of individual plants spaced out apart from each other. Filippi interprets these natural patterns in the landscape as a cue to spacing and creating rhythm within planting areas. Plants are not all grouped together in close proximity but are instead dispersed, acknowledging their competitive responses to each other as well as their own needs.

The mineral make-up of the landscape also plays a large role in defining the forms of the local plant communities and, in Filippi's view, it plays to a gardener's advantage. The stony soil makes for a free-draining porous medium that is beneficial for root systems to grow through, providing ample aeration and drainage as well as facilitating oxygen uptake through roots. The mycorrhizal networks established between roots and fungi within the substrate play an important role in providing essential elements needed by the plants. Given the paucity of organic matter in the soil, fungi and other bacteria work to dissolve the surrounding stone, releasing nutrients in a soluble form for uptake by plants.

Differing degrees of substrate composition within a site provide creative opportunities. Filippi advocates creating communities with different species by matching plants to these specific soil make-ups. This can be done through customizing different parts of the garden in different zones, in which the actual density and porosity of the soil may vary slightly to accommodate different plant communities. The zonal distribution of communities according to porosity is key, rather than any determined arrangements associated with relating individual species' characteristics to each other in aesthetic ways: 'How the plants are arranged within each zone is of secondary importance. When planning a traditional garden, gardeners often focus on precise arrangement of plants, carefully studying colours and flowering periods so that the final picture will be successful from the ornamental point of view.'[13]

His book *Bringing the Mediterranean into your Garden* (2019) details the necessity and rationale for a new aesthetic sensibility that gardeners need to adopt, which embraces the diversity of texture, colour and smell of foliage as essential elements of a planting scheme, as opposed to the overwrought insistence on flowers. Flowers for Filippi are a bonus, but morphological and textural contrasts are essential in providing rhythm and balance throughout a planted space. Evergreen plant material provides a constant structure throughout the driest months, and gives a framework that references natural plant communities; those that provide constant ground cover of different forms and heights play a primary role in determining spatial arrangement. Additionally, Filippi gives consideration to the light-reflective properties of dark evergreen leaves. In particular, sclerophyllous plants reflect different amounts of light at different times of the day, creating changing spatial effects within a garden setting. The forms of plants also play a starring role in shaping the sense of place. The archetypal cushion form of so many plants of the garrigue is an adaptation to the difficult conditions of wind, salt, cold, grazing and drought in the landscape, and can be deployed creatively within the garden (Plates 36 and 37).

Also key is working with seasonal change in a climate where the flowering period of many plants is limited to a short burst in spring followed by months of summer dormancy. Loving the yellow and brown foliage of plants over these months is something that Filippi believes needs to be learned through getting in tune with the larger landscape. In much the same way that the new perennial movement advocated leaving the skeletal structures of plants in place over winter, Filippi encourages appreciation of their decaying beauty over the hottest, driest months.

He also anticipates dynamic change over years and rejects the traditional horticultural model of gardens as 'a scene in which each plant will have its well-defined space, as in an English herbaceous border where the positioning of plants and their successive flowerings are carefully orchestrated; in a garrigue garden the plants will move from year to year, reconfiguring the original planting plan to the point where it may even become unrecognizable after a few years.'[14]

These processes of dynamic change are revealed in his view of maintenance, when faced with prolifically seeding plants that appear in places suited to their needs rather than those planned or intended by the gardener. Much like Clément, he accepts this as a natural evolution of the garden and something to which gardeners need to become reconciled: 'Instead of battling against this, it is in the gardener's interest to encourage it; by ensuring the continuity of the plantings, self-seeding reduces maintenance in the long term. In this sense a garrigue garden is a remarkable gardening school, for it invites the gardener to change roles: instead of trying to dominate nature in the garden, he or she can begin to observe the ever-changing nature that is now the daily background to life.'[15]

Fundamentally reassessing the role of the designer, Filippi believes that 'we don't want artists, we want ecologists.'[16] He stresses that designers working within this biome need to prioritize their knowledge rather than their creative skills in order to achieve sustainable schemes, and acknowledges Thomas Doxiadis, Urquijo-Kastner, Jennifer Gay and Piers Goldson, Agence APS and Bruno Demoustier as some practitioners who have taken this on board. His nursery has become a valuable resource of experimental trialling, supplier of otherwise unobtainable species, and nexus for the like-minded. In his efforts to push the focus of planting from the aesthetic to the ecological, through his website he has gathered together an informal network including designers, botanists and ecologists with the aim of sharing and increasing more information on the plants of the Mediterranean.

JAMES AND HELEN BASSON: ALGORITHMS FOR ARIDITY

James and Helen Basson are part of Filippi's local community, and are helping to push forward the science and art of planting in the South of France. Since relocating from England in the late 1990s, they have established a landscape design practice that is firmly orientated towards an ecological appreciation of the Mediterranean area. Their work has sought to put into practice sustainable, water-conscious methods of planting using many of Filippi's ideas, along with experience gained from their own exploration of the flora of the region. They see the potential of the landscape of the garrigue as a model for planting design in gardens by learning from the stress-tolerant plants and communities that inhabit it.

Like their mentor, they spend a great deal of time observing plants of the region. Reading the landscape closely provides them with an understanding of the ways in which plant communities interact functionally and aesthetically. Their projects translate the surrounding terrain into carefully considered designs that aim to ensure the ecological needs of plants are met as well as those of their clients.

An analysis of the vegetation coverage already existing around a site is essential for ensuring that the garden sits comfortably within the continuum of the environment, so plotting the spatial layout of the dominant evergreen shrubs is an important first step in the process. Using a matrix approach is less prescriptive than a traditionally specified method and achieves a more naturalistic result:

> We don't create planting plans, we create swathes of landscape, which we populate with percentages of a plant mix, and the percentage is taken from looking at the landscape. So we might walk in the landscape and see that it is 70 per cent rosemary. People might find that a bit too much, so we will break that down to maybe 40 per cent. A lot of our garden design is about how much space there is between plants, because if you are not careful you end up filling the landscape with vegetation, and when there is no lawn, there is no air, so part of the lesson we learn from looking at landscape is how much space to leave between the plants. So we might only plant 40–70 per cent of the garden.[17]

The arrangement of the structural plants is in drifts and patterns interpreted from their landscape studies (Plates 38 and 39).

While drawing on this language of the local landscape for species selection, they are considered in their deployment of the most obvious plants. Using genetic variations of a single species, such as rosemary, from around the region means that they may use as many as twenty different varieties, thereby increasing the community diversity and resilience, as well as providing a greater range of form and texture and different flowering times. They are also concerned with future-proofing their work by using species from drier parts of the region, such as Greece, Turkey and North Africa, which 'deal with severe drought, so as the landscape dries out over the next few generations, then those plants have had legs to help them cross the Mediterranean and evolve in the southern France landscape'.[18]

Planting is done in a 10-cm/4-inch gravel mulch obtained from local sources. This aids better root growth, and lower nutrient content means that plants will take longer to establish, but will be better suited to the conditions and more resilient as a result. Different grades of gravel are adopted for various microclimates, and deeper layers of mulch have been employed where they have been deemed beneficial. While importing materials to the site raises sustainability issues, they believe these are offset by the advantages of plants growing without needing any additional resources, such as irrigation, in the future.

Ensuring long-term structure requires careful selection, arrangement and aftercare, especially as many of the common Mediterranean plants, such as cistus, lavender and rosemary, are quite short-lived, 'so there is this kind of movement, a six-year rotation. There is a constant evolution and we try to ensure that the plants will evolve.'[19] Over the long term, this requires the gardener to guide the development of the plantings through selectively curating self-seeded plants as well as pruning to regulate growth and form in a manner called 'goat pruning', which emulates the grazing that takes place in the wider ecosystem.

Matching the right plant with the right place is a matter of aligning the ecological traits of plants to the site and to each other. But it is a task that exponentially increases in complexity according to the amount of information that is taken into consideration regarding both plants and the location: 'For any one area in a garden there are innumerable options as to which plant can be used where, depending on a series of variables – soil type, flowering season, origin, habitat, height, width, colour, frost resistance, speed of growth, competitive nature, shade, sun, longevity, maritime or forest to name but a few.'[20] In an unpublished essay, 'New naturalness in planting design as a mirror of nature', James Basson breaks down the factors that determine planting design into three distinct categories relating to the site, plants and cultural context.[21] The first consists of the environmental conditions, including climate, moisture availability, seasonal variation, daylight hours, soil type and gradient. The second considers plant characteristics, such as habit, mature size, growth rate, flower colour, period of flowering, texture, species origin, reproduction rate and regeneration patterns. Third are the expectations of the client: the way they wish the garden to look and be used, including its style, degrees of formality or informality, colours, seasonal interest, as well as practical considerations concerning irrigation and maintenance.

Part of the Bassons' working practice has been ascertaining ways to resolve all of these factors into a dynamic and adaptive design that balances the cultural creation of a garden with the inspiration of the natural landscape. The effect they aim to achieve is a consciously animated design in which, 'as it gets drier the species change, as it gets shadier the species change, as it gets wetter the species change, so that there is an experience of walking through landscape and evolution of the vegetation.' Using spreadsheets, they are able to plot these changes in some detail: 'At the moment that is generally what we are doing, we are varying percentages of the vegetation as we cross the landscape to simulate that evolution, but it is more an aesthetic choice. In the end what we really want to also input are all those north, south, wet, dry, steep, flat values so that it will tilt the balance of which vegetation does best in which area.'[22]

Consequently, design decisions become acts of prioritizing certain variables in order to create the desired effect: 'At some point you are selecting which columns you are using as your most important factors. Of course they are all important, and in the end it has got to grow there, but where do you start, what is your A column and your B column before you start filtering all your other ones?'[23] After analysis, the selected species are accorded percentages and distributed throughout zones within the matrix.

To deal with all these variables, the Bassons have been developing an AI program to assist them in processing the data. Having compiled a plant database, they are able to take the information on a selection of species and run it through the various parameters to come up with a potential planting arrangement for a particular site. Using these calculations helps to identify if certain plants are likely to become over-competitive, and what the ideal balance of species is to make the plant community as complex, diverse and stable as possible. Being able to process so much information opens up creative opportunities, although because of their algorithmic potential, limits are sometimes necessary: 'At the moment we also have potentially too much choice, so we often use random factors to remove some of the options. Because our problem is, as designers, we have got too many options and not enough time to choose from them, so we like to pare it down through logic, until we get to a mix of plants which is constantly different but constantly viable.'[24]

The planting mixes can then be translated into graphic form suggesting patterns that may be used when laying out some of the plants, while others are scattered on a random basis:

> The value of this approach to plant choice and layout is that it generates unpredictable solutions in each and every case, creating a more diverse range of plant communities that can be applied to different garden situations. The potential for variations within naturalistic planting provides endless combinations on an ecological and aesthetic level. The use of this model on an individual scheme, which sets its own environmental restrictions, allows the designer to deal with the vast amount of choices available in a simple way, allowing for a maximum amount of creativity within the overwhelming complexity of the natural world.[25]

The more granular the information in the system, the more detailed it can be when dealing with specific microclimates and cultural preferences. Their aim is to make the program more sophisticated by increasing the number of parameters it can use for processing. Future consideration is likely to be given to factors such as scent, an aesthetic benefit that is often a product of adaptive traits such as allelopathy, meaning that where the plant is placed will affect the other plants around it, as well as whether the garden owner can enjoy the perfume.

Taking the program online to link up with other planting databases will increase the creative possibilities and potential applications, as the more information input, the greater the scope of the calculation. Potentially, the AI architecture could accommodate plant knowledge from all the biomes, meaning it would extend its relevance to a wider international audience of designers. Becoming an open source will give it a new level of interactivity and complexity through a wider range of source material. The Bassons also intend to link it up to climate databases, which will allow for real-time evaluation of changing temperatures and precipitation levels in a specific area, more accurately aligning the information to individual species, ecological ranges and niches. This would then be able to feed back into the plant information, monitoring how plants are adapting and migrating as the climate changes and allowing better-informed decision making when selecting and planting.

PLANTS AS POSSIBILITIES

ECOTYPES: DESIGN DETAIL IN THE GENES

The permutations at play in the software developed by the Bassons highlight the myriad factors that need be taken into consideration when ensuring that plants live well in their designed habitats. The parameters are fixed by the knowledge available, and recent plant science research has been unearthing an ever increasing amount of other influences on plant characteristics and behaviour to be thrown into the mix to paint an even more complex and entangled picture of plant life.

While the majority of genetic research on plants is carried out in order to benefit agriculture by increasing yield and resistance to pathogens and environmental stresses, its application to plantings in parks and gardens opens a new frontier in which subtle investigations into provenance can enhance community resilience and function in the face of changing conditions, while still maintaining cultural aesthetic aspirations.

Although the general relationships of plants to their environments were established some time ago, it was nonetheless observed that plants within species were not uniformly consistent, but harboured certain different characteristics depending on where they were growing. In the early decades of the twentieth century, Swedish botanist Göte Turesson collected a number of hawkweed plants (*Hieracium umbellatum*) from different habitats at separate locations which exhibited phenotypic variations in morphology or physiology. Trialling them together in his garden, and subsequently at the Institute of Genetics at Åkarp, he discovered that the plants retained their variation despite being subject to uniform environmental conditions, concluding that they contained subtle genetic differences derived from adaptations to their original habitats. He used the word 'ecotype' to describe these distinct local populations of plants, which, despite their differences, were not enough to warrant them being categorized as a subspecies, and were capable of interbreeding without loss of fertility or vigour (Plate 40).[1]

Another experiment in California in the 1940s backed up Turesson's findings, and confirmed that ecotypes may occur within a specific region or be geographically distributed in widely separated places where similar conditions are prevalent. An interdisciplinary botanical team featuring geneticist Jens Clausen, taxonomist David Keck and physiologist William Hiesey, at the Carnegie Institution's laboratory at Stanford's Department of Plant Biology, conducted trials using common yarrow (*Achillea millefolium*).[2] Selections of the species were gathered from different locations along a transect line across the county stretching from the coast to the mountains. These were vegetatively propagated to provide genetically identical plants, and each variant planted in three gardens at different altitudes. The morphological and phenological differences between the plants continued unabated in each of the gardens, with the best growth and survival rates corresponding to the closest match between the altitude of their original location and that of the garden. After several other similar experiments, they concluded these types of variation occur because environmental conditions isolate populations, which then develop minor genetic adaptations that are shared by members of the population, but distinguish them from other populations of the same species.

Understanding the nuances of ecotypes in greater detail is important for ensuring a better fit of plant to place. Selecting a species that functions within a community is the first step, but then selecting the most appropriate ecotype is the next. The idea has been a topic of discussion within the field of restoration practice in the United States, which relies on using seed from local regions, known as seed transfer zones, in order to ensure fidelity with existing ecologies. This geographical approach assumes that area specific adaptations in the plant material will provide compatibility with the site from which it originates and the site in which it is to be used. Identifying ecotypes as a further step in this process can provide compliance with local character while also ensuring that there is limited but appropriately varied range diversity to prevent issues of reduced genetic variation, due to inbreeding depression from the crossing of related individuals, or outbreeding depression from crossing between two genetically distant groups or populations. Attention to site specifics such as climate, soil, latitude and altitude allows the use of a greater range of ecotypes, thus minimizing the negative consequences of establishing new plants while assuring the intended ecosystem functions.

Aside from restoring communities to an ideal target state, genetics offers a lot of scope for addressing planting in a changing climate. The fact that some ecotypes are better adapted to warmer temperatures means that they may be redeployed from their local habitats to areas that are warming and experiencing novel conditions. Different ecotypes offer different rates of carbon sequestration and oxygen provision. Variations in leaf size and shape offer alternative responses to stress factors such as light and moisture availability, with certain ecotypes better suited to a site than traditional local ones. Fine tuning by using ecotypes can extend flowering seasons, and also target the timings of different pollinator species, to the benefit of both plant and pollinator. It moves the focus from the right plant in the right place to the best plant in the right place.

Local ecotypes have traditionally co-evolved to have an advantage in their natural habitats, but their evolutionary rate of response to increasing stress factors, such as rising temperature and moisture levels, may be too slow to ensure survival. Even reductions in biomass production due to these changing conditions can lead to a reduction in carbon cycling and sequestration, which can then set off further positive feedback loops in the plant community. In rapidly transforming environments like the Arctic, this can influence the other alterations in an ecosystem, such as accelerating the melting of ice.

Ensuring that plants are genetically well suited to urban situations is a driving force behind most of the work by Henrik Sjöman, Scientific Curator at Gothenburg Botanical Garden. Beginning with his doctoral thesis in 2012, he has been applying the idea of fine-tuning plant selection by considering different ecotypes for trees in order to 'develop and test a working procedure for identification of new tree species and genotypes that holds the potential to diversify urban tree populations'.[3] Appreciating the important functions trees play ecologically in cities, Sjöman's concern is with ensuring that trees are appropriately suited to their sites not simply for their establishment when they are planted, but also for ensuring their long-term success facing the vicissitudes of the future. His research with colleagues into drought tolerance for trees has shown 'a significant relationship between the potential evapotranspiration of the provenance collection site and . . . significant positive differences in drought tolerance between provenances and subspecies . . . By directing efforts towards identifying more drought-tolerant genotypes, it will be possible to diversify the palette of trees that could confidently be integrated by urban tree planners and landscape architects into the urban landscape.'[4]

Sjöman has been addressing the issue of selection in creating novel plant communities as part of a research collaboration on future flora between the University of Sheffield and the Swedish University of Agricultural Science, with a team including James Hitchmough. They have been delving deeper into the idea of genotypically adapted planting in urban areas. While the importance of using plants in densely built environments is now widely accepted, it means that efforts to get more of them into cities increasingly need to consider how they are used in restricted, moisture deficient sites often bound by hard surfaces, like roadsides, pavements and traffic islands. The research acknowledged that many traditional herbaceous species are simply not suited to such environments, and undertook a reconnaissance of similar types of plants that exhibited suitable traits from natural habitats in Europe. Fieldwork has gathered 'knowledge on the growth and performance of the different species/genotypes in another climate region and their tolerance towards local biotic and abiotic stress agents . . . a genetic pool of available species, from which to select types fit for purpose. This includes differences in growth, leaves, flowering, etc.'[5]

In another report about selecting trees for creating urban forests by Sjöman and Hitchmough, with Harry Watkins and Ross Cameron from the University of Sheffield, the authors suggest that using information from 'biogeographical studies would require a fundamentally different approach to species selection, requiring urban foresters to understand and harness evolutionary adaptations, target specific populations or ecotypes and then match these to specific designed environments. Such an approach would enable a far greater degree of precision and confidence in designing urban forests to meet specific challenges.'[6] The knowledge base for doing this is currently the limiting factor, as most 'practitioners rely upon specialist horticultural texts (the heuristic literature) to inform species selection whilst the majority of research is grounded in trait-based investigations into plant physiology (the experimental literature). However, both of these literature types have shortcomings: the experimental literature only addresses a small proportion of the plants that practitioners might be interested in whilst the data in the heuristic (obtained through practice) literature tends to be either too general or inconsistent.'[7]

In order to ensure that this can be done practically within the nursery and design industries, Sjöman and Watkins have been investigating to what extent provenance and ecotypes are understood within the largest tree nurseries in Germany, the Netherlands and the UK. Setting out their manifesto, they suggest, 'it is not currently possible for specifiers to select trees at an intra-specific level based on climate or ecological criteria. If the goal of urban forestry is to create long-term sustainable tree populations that develop large, prosperous trees over time, it is of the utmost importance that the plant material that is used is of the best possible fit with the target site. This means that the plant material is of a genetic origin that is ecologically suited to both the climate and growing conditions on site.'[8]

BEYOND COMPETITION: INTIMATE ACTS OF FACILITATION

The history of plant community science has, along with the rest of ecology, been largely driven by a focus on the more negative influences on plants, such as competition, stress, disturbance and predation. In particular, ecological planting has placed a high premium on competition and stress as the key drivers that define the structure and diversity of plant communities, as outlined in Grime's CSR model (see 'Plants as Processes'). Consequently, plants are considered in terms of their vulnerability to the vicissitudes of abiotic factors, as well as predators, pathogens and other plants. While there is empirical evidence of these determinants, often they can overshadow the influence of other contributing interactions, in particular those of a more collaborative or cooperative kind.

This is characteristic of ecology as a whole, which ever since Darwin has adopted a perspective based on conflict and privation, and their effect in removing species and depleting communities. This has more to do with an interpretation of Darwin's theory of evolution promoted by one of his followers rather than the wider perspective he actually proposed. Herbert Spencer popularized the idea of 'survival of the fittest,' suggesting that only the strongest organisms survive, as opposed to Darwin's intention that the organisms best adapted to their environment survive and evolve. He suggested that they achieve this by both positive and negative interactions with biotic and abiotic forces.

The simplistically aggressive model has sometimes clouded a clearer vision of inter-species relationships, and certainly created a dubious meme that has infected social ideologies up to the present day. The notion in the natural sciences – of individuals always in a conflictual mode for dominance – developed largely within the framework of Western post-Enlightenment individualism, reinforced by practitioners who were predominantly men, which gave it an unconscious gender bias towards aggressive, competitive behaviour despite the assumed neutrality of scientific discourse.

Also, anthropomorphizing plants by interpreting their behaviour as conforming to the same actions as humans can bring with it loaded preconceptions, such as when gardeners start talking about thugs in their borders. An interpretative framework is obviously necessary, and many behavioural traits can be seen to be similar across species, but framing everything in the world around humans can downplay the importance of other species-unique characteristics.

However, since the 1990s there has been a growing focus on how more cooperative interactions play important roles in plant communities. Positive interactions between plants, known as 'facilitation,' are relationships that are beneficial to at least one of the parties involved and detrimental to neither. Cases where both parties benefit are called 'mutualisms', and 'commensalism' is when the effects of the facilitated on the facilitator are neutral. These interactions have been found to be globally ubiquitous, occurring in all type of plant communities, and to vary across a scale of magnitudes, including even transient interactions with small effects, as long as there is a net positive effect. Some are the result of specialist co-evolution between two species, while others are more generalist interactions.

Facilitation occurs when one species alters the local environment for another species, such that one enhances the other's prospects of growth, reproduction and survival. They may do this directly by reducing temperature, moisture or nutrient stress factors, which can simply be an effect of their presence in a community, phenology and life cycle. Trees do this by moderating the understorey light and water conditions, as well as contributing to nutrient levels through leaf litter. Facilitation can also be indirect, such as when a facilitating plant may remove another competitor, improving the situation for the facilitated plant, or when it deters predators. These interactions may take on another level of sophistication when the facilitator interacts not directly with a competitor but on another competitor, which then affects the first one.

Some plants will provide shelter and create conducive conditions for others in harsh situations. In arid environments, shrubs can act as 'nurse plants' facilitating amenable circumstances for the germination of herbaceous species seedlings by providing shelter from wind, reducing temperature, and regulating moisture and humidity. But while these 'Goldilocks' situations provide amenable conditions for germination, they can also negatively affect juvenile growth by casting shade and reducing light, highlighting the interplay between facilitation and competition that can occur, either simultaneously or over a period of time. Facilitation therefore has the ability to affect spatial arrangement of communities and the possible locations of various species. It can expand the realized niche of a plant even beyond the limits of its fundamental niche. This means a certain species can endure stress that it would not normally be able to cope with. These interactions could provide survival strategies for many species when facing a changing climate.

This adds more levels of complexity to plant communities, especially when these positive interactions, along with the negative ones, are occurring between many different species at different times and across different life cycle stages. Direct and indirect facilitations need to be considered, taking into account potential feedback effects. Research has focused on defining the mechanisms and results of the interactions in terms of individual growth and fitness, spatial distribution of populations, and the composition, diversity and dynamics of communities. The information garnered could be integrated into competitive models to consider the balances between them. It can also help to determine the stress toleration, as facilitation is more common in high-stress environments and competition in low.

Knowing the balance between positive and negative interactions is important, and requires knowledge of both the facilitator and the facilitated, and an understanding of their co-evolutionary histories. An awareness of the idiosyncrasies that facilitation produces can add to existing plant community analysis, to build a bigger picture of all the dynamic forces and gradients they occupy within an environment which affect plant-to-plant relationships. It offers environmental benefits that can respond to a changing climate, as well as creative planting possibilities for establishing novel plant communities.

THE RHIZOSPHERE: THE UNSUNG UNDERGROUND

The results of relationships such as competition and facilitation are expressed physically above ground in a visible manner. The effects of plants casting shade, spreading clonally, seeding prolifically, and crowding out other neighbouring species can easily be seen. However, some of the most important mutualisms are actually happening underground, out of sight, and it is easy to understand how they may have been missed as important factors in plant community constitution. But mounting research is confirming that incredible forms of interactions within the soil are quite as important as those happening above ground.

Soil quality has long been understood as an important factor in the health of plants, and ecological planting has recognized that, as a limiting resource, either high- or low-nutrient availability plays a role in both individual plant development and community structure. It can facilitate the prevalence of competitor or stress tolerant species, and help to determine community composition and physiology. But the influence of soil is due to more than its chemical and mineral properties as generally described in horticulture, and recently the secrets of the soil have revealed a lot more about the complex mixture of biotic and abiotic elements that affect plant behaviour and community assemblage.

Soil is a mixture of organic matter, minerals, liquids and gases, and the narrow layer just below the ground surface, which is home to plant roots, is an incredibly diverse and dynamic realm known as the 'rhizosphere', a term coined in 1904 by the German agronomist and plant physiologist Lorenz Hiltner to describe the soil-to-root interface.[1] As essential parts of plants, roots not only provide anchorage but are the underground portion of the system that transports carbon – absorbed during photosynthesis from the atmosphere – down to the soil, and conversely conveys extracted nutrients and water from the soil to fuel plant growth and reproduction. The rhizosphere houses an impressive amount of subterranean plant biomass featuring a diverse range of root morphologies that allow an effective use of space for the different species that comprise plant communities, in a similar way that the architecture of stems, branches and foliage does above ground. The balance between the above- and below-ground parts of plants is expressed in root-to-shoot ratios, which vary according to different types of plant communities. Grassland habitats have more root biomass than forests, reflecting various evolutionary responses to environmental conditions. The ratio is negatively correlated with mean annual temperature, precipitation, plant height and shoot biomass, but positively correlated with elevation and latitude.[2]

The most important aspect of the rhizosphere is that it is the living biome with the greatest concentration of life forms on the planet. Most noticeably, the visible soil-dwelling invertebrates such as worms, ants and termites have long been considered as underground engineers, actively engaged in physical, chemical and biological processes such as aeration and nutrient cycling. Microfauna such as nematodes and small arthropods play further roles as decomposers and food sources on lower trophic levels. But the most important part of the rhizosphere is the extremely dynamic and diverse living microbiome consisting of an incredible array of micro-organisms, up to one billion per teaspoon of soil, including bacteria, fungi, protozoa, algae and archaea, which work together in complex ways to influence biogeochemical cycles, ecosystem structures and functions, and more specifically have both beneficial and detrimental effects on plant growth and health. The complex interactions between the plant, soil and microbes can be antagonistic, mutualistic or synergistic, according to the specific associations between them.

The range of soil microbes includes many phylogenetic groups, showcasing an incredible amount of genetic diversity, with up to several thousand different genomes per gram of soil, and functionally consisting of producers, consumers, and decomposers, which play essential roles in sustaining plant growth through such functions as water purification, atmospheric nitrogen fixing, and mobilizing and solubilizing organic and inorganic nutrients. Collectively, they are often considered as the second genome of the plant and are crucial for plant health, with their industrious endeavours fulfilling most of the nutritional needs of terrestrial vegetation. Both aerobic and anaerobic bacteria and fungi use chemical structures and substrates to break down plant litter and microbial necromass, to liberate carbon, nitrogen and minerals, making them available for plants. They affect all stages of a plant's life cycle from aiding seed germination to enhancing seedling vigour and plant growth, as well as preventing the threat of pathogens. The provision of carbon-containing compounds from plant roots, known as rhizodeposits (originating from exudates, sloughed-off cells and mucilage), provide microbial sustenance, regulate the diversity of micro-organisms, and influence their symbiotic activity on roots.

The microbial populations assist not only in ensuring plant fitness, but also in determining adaptive traits and mechanisms, and two distinct types of association have been observed.[3] The first is microbe-mediated local adaptations, which occur when a plant has a genetic affiliation with specific microbes in a local environment. Based upon an evolutionary tendency, the plant attracts, retains and regulates important microbes to its advantage in obtaining resources and adapting to soil conditions. The other is microbe-mediated adaptive plasticity, which is a product of a plant's ability to thrive in a variety of environments because it is plastic enough to adapt and make interactions with the various microbes in each particular environment, thereby benefiting from resources and increasing its potential range for stress tolerance. Both of these modes give certain plants a survival advantage over other plants which lack either the suitable genetic affiliations or plasticity. These advantages are then key factors in how plants associate with each other, affecting community composition.

Further to the different strategies of association, it has also been discovered that roots are not the only part of the plant to play a part in interactions with microbes. Another aspect of the relationship concerns the process of decomposing, dead-plant biomass, in the forms of root and leaf litter, which is one of the most important ways in which energy and carbon are transformed and cycled through terrestrial ecosystems. While many decomposers are likely to be generalists, their performance when assembled into specific trophic communities may be directly linked to the host plant species. The Home-Field Advantage (HFA) hypothesis suggests that because litter from different plants varies in physical and chemical characteristics, it has consequences for the make-up of microbial communities and decomposition rates.[4] Research has shown that this is particularly apparent at larger scales such as forests, where localized adaptations between trees and micro-organisms create a home-style environment. Decomposition has been observed to be quicker at home, by up to 10 per cent faster, than away in other habitats with different species. This consequently affects nutrient release, with knock-on effects on plant growth and fitness, competitive relationships within soil communities, and rates of succession.

HFA has been primarily shown to exist in environments with dominant species present, such as forest habitats, yet there have been further proposals that postulate that HFA could also exist within mixed-plant communities operating at the level of individual plants. This suggests that home environments may exist at the micro scale as well, with specific soil webs that differ from the overall mean community aggregate. These nuanced distinctions may well be drivers of generally observed plant characteristics and behaviour, but accounting for them along with other more obvious biotic and abiotic factors could provide greater understanding of the balance between individual plants and community structures. The insights of HFA, however, are just a part of a fascinating bigger picture of reciprocal interactions taking place above and below ground which shape the environment.

MUTUALISMS

While research into facilitation found a firm footing at the end of the twentieth century, the hidden history of mutualisms was already well underway, and had been providing an alternative narrative to the one constructed around theories of competition. In 1867, Swiss botanist Simon Schwendener proposed a hypothesis that lichen was formed by two separate organisms, a fungus and an alga, an idea confirmed a decade later by German botanist Albert Bernhard Frank, who proposed the term 'Symbiotismus' or symbiosis, meaning 'living together' in Greek, for this unique union. His term described the relationship between two genetically different organisms, regardless of how casual or essential, distant or intimate the bond was, or whether it had positive or negative effects on either party. The idea was then popularized shortly after, in the monograph *Die Erscheinung der Symbiose* (1879) by a German botanist and pioneer in plant pathology, Heinrich Anton de Bary. Frank then dug deeper in his 1885 book, *Über die auf Wurzelsymbiose beruhende Ernährung gewisser Bäume durch unterirdische* (About the root symbiosis-based nutrition of certain trees by underground fungi), exposing symbiotic interactions between plant roots and fungi in what he named 'mycorrhiza', meaning 'fungus-root'.

This dualistic relationship is the result of co-evolution between the majority of land plants and mycorrhizal fungi (Plate 41), which began over 450 million years ago, and is believed to have played a central role in the migration of plants from aquatic habitats to terrestrial ones. As a result, around 90 per cent of plant species are reliant on mycorrhizal fungi. They fall into two main groups: 'ectomycorrhizal', which are external to roots and form a mycelial sleeve around them, and arbuscular or 'endomycorrhizal', which penetrate into the cells of the root. Ectomycorrhizal fungi are a diverse group of about 10,000 known species, yet occur less frequently – in around 2 per cent of plant species. They tend to have co-evolved unique bonds with certain species, usually woody plants from the birch, dipterocarp, myrtle, beech, willow, pine and rose families. Around 80 per cent of mycorrhizae are abuscular, featuring 150–200 known species, and are non-host specific. This allows one fungus to be able to connect simultaneously with multiple plants, including those at different life cycle stages, from seedlings to mature plants, and across different species.

The relationship between plants and fungi is based upon a reciprocal exchange of metabolic resources. The fungi provide the plants with nutrients, including up to 80 per cent of their nitrogen requirements and 70 per cent of phosphorus, as well as sulphur and micronutrients, which they acquire from the soil through foraging hyphal networks. Apart from supplying nutrients, other benefits include regulating pathogens, influencing photosynthetic activity and excluding toxic ions. In return, plants provide the fungi with sugars and lipids produced during photosynthesis, with arbuscular mycorrhizal associated plants allocating between 4 and 20 per cent more photosynthates to mycorrhizal roots than to non-mycorrhizal roots, and ectomycorrhizal associated plants supplying over two times more carbon to their roots than to non-mycorrhizal ones (Plates 42 and 44).

Photosynthate allocation has been observed to vary across biomes, with the least distribution happening in tropical forests and the most in grassland and tundra biomes. The expense to plants of distributing these photosynthates to fungi is offset by the fact that it is less costly for them than to grow their own roots, as hyphae grow more quickly, can penetrate more easily through smaller pores in the soil (as they are up to fifty times finer), and create more expansive networks given that they are around one hundred times longer. The mutualistic relationships between plants and fungi are not simply metabolic interactions but can also assist in alleviating environmental stress, influence the adaptive behaviour of plants, and facilitate rapid changes in plant physiology, gene regulation and defence responses.

There can also be more complex processes between multiple plants and mycorrhizae, which can produce effects at a community level, such as when carbon in forests is transferred from overstorey foliage to plants in the understorey deprived of light and unable to photosynthesize as efficiently. Bidirectional carbon transfers between ectomycorrhizal tree species were described on the cover of the August 1997 issue of the journal *Nature* as 'the wood-wide-web', in relation to the experiments into hyphal networks between different species of trees by Canadian forest ecologist Suzanne Simard. Building on research from the 1980s by plant biologist David Read of the University of Sheffield, Simard's work has opened up a fascinating new perspective on the life cycles of trees as interconnected and interdependent, using ectomycorrhizal networks to communicate signals and cues between each other in the form of chemical substances in order to affect behaviour and trigger morphological and physiological changes as responses to environmental challenges.[1] She has revealed previously unknown levels of behaviour: 'Mycorrhizal fungal networks linking the roots of trees in forests are increasingly recognized to facilitate inter-tree communication via resource, defence, and kin recognition signaling and thereby influence the sophisticated behavior of neighbors. These tree behaviors have cognitive qualities, including capabilities in perception, learning, and memory, and they influence plant traits indicative of fitness.'[2]

She has gone further to suggest sophisticated supportive relationships, interpreted in kinship terms, with 'mother trees' nurturing offspring:

> Some trees shuttle allelochemicals, or poisons, through the network if the neighboring tree species is an unwanted intruder. Elder trees are able to recognize neighbors that are genetically related, or that are kin, and they can send more or less resources to other trees to either favor or disfavor them, depending on the safety of the environment. I have taken to calling these elders 'Mother Trees' because they appear to be nurturing their young. Mother Trees thus connect the forest through space and time, just like elders connect human families across generations.[3]

More generally, the dynamics of the community-level effects of symbiosis have proved to be an important field of research investigating the mechanisms between plant and microbial communities that determine the assembly and productivity in each. Increases in fungi have been shown to be related to increases in plant biomass and diversity in communities. Philip Grime, working with David Read in 1987, framed the latter's fungal findings as another contributing factor influencing community composition, additional to the standard CSR determinants, recognizing that 'mycorrhizas increased diversity markedly by raising the biomass of the subordinate species relative to that of the canopy dominant . . . Export of assimilate from canopy to subordinate species through a common mycelial network is likely, together with enhancement of mineral nutrient capture, to be involved in the beneficial effect of mycorrhizas.'[4]

Further research has attempted to ascertain the mechanisms behind these phenomena, and a number of conceptual approaches have been applied in trial situations. These interactions are understood as plant–soil feedback, in which plants affect microbe communities and, conversely, microbes are active in determining the composition of plant communities, in both positive and negative ways. Empirical results have indicated that negative feedbacks create or maintain diversity within the community, while positive ones result in a decrease through the creation of competitor partnerships.

The main debate, however, surrounds the question of whether one community is instrumental in initiating the feedback rather than the other. The 'Driver' hypothesis suggests that, at a community level, changes in fungal communities can direct changes in the vegetation, while conversely in the 'Passenger' hypothesis, changes in the vegetation regulate fungal assemblages, recruiting and selecting particular fungi as partners. Another theory, the 'Habitat' hypothesis, throws into the mix the idea that the environment shapes both communities, assuming that their interdependence is due to their independent adaptation to the same abiotic conditions, thereby causing correlated variation.[5] This appears to apply to the bigger perspective of the spatial distribution of communities across a range of different environments rather than within environments in similar conditions. It may be that all of these could be occurring simultaneously at varying scales and operating independently in localized conditions, giving rise to multidirectional effects and making the picture of their interrelationships as tangled as the mycorrhizal networks themselves.

Adding in all the other microbial forms of life in the rhizosphere, and accounting for the roles they may play in feedback with plants, fungi and each other, build in yet another layer of complexity to community functions, both below and above ground. Aside from the general directional influences, more granular detail remains to be explored, to discover whether there are specific micro-organisms that play more influential roles than others in generating either positive or negative feedback, or certain conditions that produce beneficial or antagonistic associations. Genetic sequencing is leading the way.

The additional factor of a changing climate gives rise to concerns about future plant–soil feedback. Rises in intensity and abundance of extreme conditions like drought and flooding will affect not only plants and fungi individually but also their symbiotic relationship. Given that the latter are instrumental in maintaining soil stability and enabling carbon retention, this is likely to cause feedback scenarios on a larger environmental scale. Drought reduces the rate of microbial functions and abundance, and affects certain susceptible species more than others. It also changes the balance between fungi and bacteria, reducing the importance of activities of the former and increasing the latter within the soil web, which is likely to affect plant diversity and abundance.

Appreciating the inherent complexity of these relationships within and between plant, fungal and microbial communities is fundamental to progressing towards a more ecologically integrated approach to plant communities, and developing ways in which this information may be applied within planting design. The addition of these symbioses to the traditional framework of biotic and abiotic factors that influence plants throughout their life cycles can facilitate a more nuanced understanding of individual plants and communities. The idea of 'right plant, right place' really needs to also accommodate 'right microbes'. This is certainly a bold step beyond the floral-focused fascinations of planting in the last century, but given developing technology and the ability for on-site analysis in the future, it could be an important part of an expanded strategy for ensuring a closer fit between plant and place which enhances diversity and function above and below ground.

If healthy microbes make healthy plants, then gardening should be an act of stewardship of the soil, minimizing disturbance below ground and nurturing the microbial life therein as much as tending to the plants they support. It brings into focus questions regarding the effects of anthropogenic interventions. What are the consequences of every intrusive act that breaks the surface of the soil? Is sowing seed less disruptive to microbial communities than planting? As future analysis reveals more about these life forms and their behaviours (along with more detail about their symbiotic interactions with plants), the ways in which we create and manage planted environments could be given a new grounding.

OLD FRIENDS AND NEW ALLIES

Microbes are not only crucial for roots below the ground but are also essential inhabitants of other parts of plants, such as the phyllosphere, consisting of the external surfaces of stems, leaves, flowers and fruits. This harbours a rich assortment of various species of bacteria, fungi, algae, archaea and viruses, with up to 10 million microbes per sq cm/64.5 million microbes per sq inch of leaf surface, all metabolically attuned to their unique environment. These epiphytic bacterial communities are assembled according to environmental and host-specific determinants. While they share some commonalities with rhizosphere communities, few parallels seem to exist with airborne organisms. While not as diverse as the phyllosphere, the endosphere hosts micro-organisms within a plant's internal tissues, predominantly arbuscular mycorrhizae and other endophytic fungi, which may differ in the above and below ground parts of the plant. Their functions span from mutualism to pathogenicity, while their diversity and composition depend on the plant species, the organ they inhabit and its physiological conditions, plant growth stage and the environment. The cellular count of micro-organisms, both on and within plants, can even exceed the number of its own cells, operating alongside the genome, and helping to determine specific traits and behaviour. These collective plant microbiomes, along with the host plant, form a 'holobiont' (a term derived from the Greek words for 'whole' and 'life'), an idea initially defined by Lynn Margulis and René Fester in their 1991 book, *Symbiosis as a Source of Evolutionary Innovation* (Plate 43).

Just as plants are inextricably bound to microbial life, so are humans in similar ways. We are hosts to billions of resident and temporary microbes, which contribute to making us the species that we are. Over half the cells of the human body are actually those of bacteria, viruses, archaea and micro-eukaryotes. The importance of these symbioses can be seen in our species history, in the evolution of mitochondria from bacteria, as well as through horizontal gene transfers with viruses. These organisms perform a variety of indispensable activities that enable us to operate at the most basic levels, from gut flora's role in digestion through to oral microbes providing resistance to pathogens. Their abundance and influence on our activities call into question the notion of what a human being actually is, as well as the idea of the individual self inherent in Western Enlightenment thinking. It is becoming increasingly apparent that interspecies microbial exchanges are continuously taking place within ecosystems, and that the microbes associated with plants can facilitate not only improved functioning in environments, but also health in humans, which opens up exciting possibilities for designed spaces that can leverage the reciprocal relationships between them.

The Microbiome Rewilding Hypothesis, proposed by Jacob Mills and colleagues in 2017, is an idea premised upon the fact that humans co-evolved in symbiotic relationships with micro-organisms in biodiverse natural environments that provided immune protection from pathogens, but due to increasing urbanization and separation from these types of habitats, the protections they previously afforded have been lost. Given that exposure to these natural environments produces an appropriate succession of gut microflora and stable immunoregulatory development, they suggest that 'restoring biodiverse habitats in urban green spaces can rewild the environmental microbiome to a state that enhances primary prevention of human disease.'[1] The benefits of this are twofold: increasing biodiversity, particularly with plants as the central feature, offers positive environmental gains through improved ecological functions, especially in impoverished urban areas, and it also improves human health. The links between environment and health can potentially reframe and drive new policy approaches, linking the two together, and given enough political will, funding across the sectors can work synergistically rather than in silos.

The air provides an interface between soil, leaves and humans in which microbes can circulate, offering situations that can result in increased skin and nasal microbial diversity and altered human microbiota composition. These aeroplankton are a mix of around 1,000 different species of bacteria, around 40,000 varieties of fungi, and hundreds of species of protists, algae, mosses and liverworts; many of them are thought to come from soil and plants, originating in the rhizosphere and phyllosphere. Due to their size and weight, they are readily transported by air currents and form vertically layered aerobiomes, more abundant and diverse at lower levels than at higher ones. This opens up questions of engagement between people and plants within urban environments. Biodiverse environments have been observed to offer greater potential for health benefits, but identifying exactly which spaces provide the most advantages for airborne vectors is key. Research suggests that more naturalistic, diversely planted spaces offer more than highly manicured, monocultural ones like lawns, and that 'wildscapes', composed of spontaneous vegetation in urban areas, also have a lot to offer.

Research by Jake Robinson of the University of Sheffield and international colleagues suggests that 'there could be value in determining whether different habitats and vegetation management regimes impact vertical stratification in urban green spaces, and elucidating the downstream health effects on urban dwellers,'[2] and that there are 'prospective landscape and social interventions that have the potential to enhance our connections with the natural world, through health-inducing microbial interactions and psychosocial pathways'.[3] Therefore the possibility of facilitating beneficial microbial exchanges between plants and humans becomes something that could be the result of conscious decision-making processes. This is summed up in the idea of microbiome-inspired green infrastructure (MIGI), a collective term used by Robinson, Mills and Breed 'for the design and management of innovative living urban features that could potentially enhance public health via health-inducing microbial interactions'.[4] Optimizing opportunities for transmission becomes a design issue for planted spaces in urban areas, with layouts ensuring proximity to soil and plants, and provision of seating or possibilities for various activities, which all increase dwelling time and therefore potential exposure, with particular consideration given to the vertical levels of the airborne communities (Plate 45).

Creating plant communities based on microbial relationships which transfer organisms between different spheres, providing both environmental and health benefits, would certainly tick a lot of boxes. This is a far cry from the colourists' ploys of planting in harmonies, tones and tasteful floral contrasts, and opens up new creative possibilities for marrying form and function, aesthetics and processes, in planting design. Taking these ideas further, assuming that advantageous microbes can be associated with specific plants, then planting schemes could in effect be cocktails of beneficial bacteria awaiting transmission to people as they enjoy gardens and public parks.

While the health benefits of green spaces have been shown to be psychologically restorative, and even related to the neurological process, the holobiont approach offers a more fundamental genetic basis. Drawing on all the interactions between plants and micro-organisms within the rhizosphere, phyllosphere, endosphere and aerobiomes, the future becomes a symbiocene in which gardens and planting can be created to be quite literally and physically good for personal and planetary welfare.

BIOCENOLOGY

To make things even more complex, microbe relationships with plants also affect plant relationships with other more visible species above ground, such as pollinators and herbivores. The positive and negative interactions of plants with animals and invertebrates play critical roles in shaping plant communities. The dependence on pollinators for reproduction has obvious consequences for survival, as do the disturbances produced by herbivores, which potentially threaten the immediate existence of plants. Many of these are apparent to both the delight and annoyance of gardeners, and the fascination with wildlife gardens touches on some of these aspects of plant life, by encouraging planting that attracts other species, but the interactions are usually more attuned to anthropocentric feelings of doing good, rather than anything strategic to enhance the greater ecosystem for the benefit of all parties involved.

Expanding the plant community to integrate these other life forms and the relationships between them is covered by a field of research called biocenology, a mixture of plant science and zoology (Plate 46). The basic unit of study is the biocenose, a term introduced by German biologist Karl Möbius in 1877 to describe a community of different species within a specific area which share the environment based upon their adapted suitability to its environmental conditions. Within this community are a variety of mutually beneficial and detrimental interactions, many based on nutritional chains. Some of these species have specialized co-evolved relationships while others are generalist, and some of the participant species need not always be present in the community at different stages in their life cycle, or may only be present seasonally or even accidentally.

Knowledge of the possible permutations of interactions between plants and other species helps to build a bigger and more complex picture of how plants live and how they function within communities. Many of these relationships are crucial for plant communities, or planting schemes, whether they be horticulturally or ecologically orientated. This adds another layer of complexity to planting design. If consideration of plant communities includes the abiotic and biotic factors of the environment, microclimatic differences within the site, and also the interactions between plants and other species, the bigger picture that emerges is a more integrated and complex one. The success of planting may be make or break depending upon these interactions, and research from biocenology will no doubt play a greater role in the process in the future as species migrate at different rates, looking at the adaptive changes that will result, and the ways in which they affect co-dependent relationships.

PLANT INTELLIGENCE

Plants' ability to respond to their environments and each other – and as a result thrive in complex ecological systems – is testament to their evolutionary traits and strategies. Yet while it is known that they do these things, less is understood about the actual means they use. Plants respond to external triggers, and their different abilities to process this sensory information range from the simplistic to the complex. Research into the ways they do this has developed into the new, and often controversial, scientific field of plant intelligence.

Experience-based behaviour is taken for granted in humans and animals. Memories of cause and effect provide a pre-emptive basis for making judgements that will, at the most basic level, be helpful for survival. Yet sentience of this kind has generally not been considered an attribute of plants. Although plant science shows that plants actively contribute to shaping their environments, this is seen as a co-evolutionary characteristic of their functioning. Their sessile nature is often seen as stripping them of the ability to be fully active participants in their environments. As passive respondents, plants are observed to be physically reacting to stress, disturbance and threats simply as defensive and recuperative actions. Yet there is a growing body of evidence to show that there is a lot more going on than meets the eye.

The idea that plants possessed some type of intelligence, which allowed them to experience stimuli and activate suitable responses, dates back to the nineteenth century. Charles Darwin noted plants' ability to communicate with their environment, and to use this information to direct the movements of their organs, in his 1880 book, *The Power of Movement in Plants*, but he failed to pursue the idea with further research. Indian biologist Jagadish Chandra Bose conducted experiments in the early twentieth century into the electrical nature of the conduction of various stimuli in plants, and developed the idea of playing music to them. Inspired by Bose, ex-CIA interrogation specialist Cleve Backster put his polygraph test skills to work on plants in the 1960s. From his experiments he deduced that plants had a telepathic ability to understand human intentions, something he termed 'Primary Perception'. Reactions from the scientific community acknowledged that electrical current flowing through plants could be subject to external manipulation but refuted the parapsychological claims. Backster's ideas were disseminated to a wide audience in a surprisingly popular book published in 1973, entitled *The Secret Life of Plants*, a sensational exposé of plant perception by Peter Tompkins and Christopher Bird, which promoted the idea that plants had emotional lives that originated from a supernatural realm beyond the bounds of science. This rocky start to the idea of plant intelligence doomed many subsequent investigations into the matter to dismissal by the scientific community and ridicule in the popular media. The idea of talking to plants became a derogatory meme, demeaning both the talker and the talked-to, relegating plants back to their lowly status in the hierarchy of being.

Despite these dubious beginnings, the nascent science has recently been gaining ground through much more acceptable means of empirical research and theoretical proposition. An article in *Nature* in 2002 by plant biologist Anthony Trewavas, 'Plant intelligence: mindless mastery', proposed that in evolutionary trade-offs plants eschewed movement in return for not only physiological characteristics but also the ability to coordinate responses to external stimuli and the means to communicate. Trewavas wrote, 'Plants continuously screen at least 15 different environmental variables with remarkable sensitivity – a footprint on the soil or a local stone, for example, are perceived and acted upon. We either know or can guess the receptors for most of these signals, which are transduced in fractions of a second through large numbers of small GTPases [enzymes], second messengers and a thousand protein kinases. The flow of information is continuous. Integrated responses are constructed after reference to the bank of internal information that specifies the plant's ecological niche.'[1] Internal communication of information between cells and tissues was ascribed to the flow of 'proteins; nucleic acids; many hormones; mineral, chemical, hydraulic, mechanical, oxidative and electrical signals; peptides; various lipids; sugars; wall fragments; and other complex carbohydrates. Quite how individual plant cells accommodate this prodigious amount of information is not understood. But even anatomically uniform cells exhibit enormously different responses to a single signal. A huge reservoir of individual cell behaviours can be coordinated to produce many varieties of organism behaviour.'[2]

The resulting behaviour determines the ways in which plants avoid competitive neighbours, find optimal light, forage for space, and evaluate humidity and mineral gradients in the soil. Trewavas found all this unsurprising as plants had 'evolved to optimise fitness. Plants must then have access to an internal memory that specifies the optimal ecological niche in which maximal fitness, usually regarded as greatest numbers of viable seeds, can be achieved. When the niche is sub-optimal, plasticity in growth and development intervenes to counterbalance and to attempt to recover, as far as possible, the benefits of the optimal niche.' In order to do this, the plant 'requires two things: (1) a goal (or set point), usually determined in advance, and (2) an error-indicating mechanism that quantifies how close newly changed behaviour approaches that goal.'[3] Trewavas was keen to stress that this form of learning-based intelligence was an extension of existing autecological knowledge about plants, and not to be confused with human intelligence or the notion of supposed free will, a confusion that has plagued much criticism of the subject.

The ways that plants perceive their circumstances and respond to environmental inputs in an integrated fashion was dubbed 'Plant Neurobiology' by Italian plant physiologist Stefano Mancuso at the First Symposium on Plant Neurobiology in 2005. Drawing parallels with animals, he and colleagues suggested that plant root apices act as command centres, and 'vascular elements allow the rapid spread of hydraulic signals and classical action potentials resembling nerves.'[4] The symposium corralled together the prime movers in the discipline and provided a round-up of the knowledge at the time. Topics investigated included plant physiology, the roles of pheromones and volatile organic compounds to attract and repel friends and foes, neurotransmitters, signalling pathways and electrophysiology.

While these biological processes continued to be explored over the following few years, the notion of plant communication was highlighted by Slovak cell biologist František Baluška in a number of edited research anthologies. The question at the heart of these books was if it was possible that there was something resembling a language structure involved in interactions between plants and other species, and, if so, how it would operate. Baluška explained that 'biocommunication in plants integrates both biology of plants and communicative competences of plants. It allows more coherent explanation and description of full range of behavioural capabilities of plants that cannot be covered by mechanistic or even reductionistic approaches. Natural communication assembles [the] full range of signal-mediated interactions that are necessary to organize coordinations within and between cells, tissues, organs and organisms.'[5]

Outside of academia, the public began to get insights into this intriguing world. Plant geneticist Daniel Chamovitz's 2102 popular science book, *What a Plant Knows: A Field Guide to the Senses of your Garden – and Beyond*, presented an enthusiastic yet cautious picture; he was hesitant about the linguistic use of neurobiology and intelligence, instead preferring to talk of plant awareness. Using anthropomorphic analogies, Chamowitz outlined how plants see, smell and feel as well as know their location and remember. His sensory summation of plant life delved into the mechanism of photoreceptors to maximize light efficiency, volatile compound scents, touch activated genes, remedial responses to stress and transgenerational gene transfer. Despite the use of metaphors for reference, his overview of plant life very much downplayed the connotations that such similarities between plants and other species may inspire:

> A plant is aware of its environment, and people are part of this environment. But it's not aware of the myriad gardeners and plant biologists who develop what they consider to be personal relationships with their plants . . . the flow of meaning is unidirectional . . . These terms represent our own subjective assessment of a plant's decidedly unemotional physiological status . . . We project on plants our emotional load and assume that a flower in full bloom is happier than a wilting one. If 'happy' can be defined as an 'optimal physiological state', then perhaps the term fits.[6]

Soon after, in a more proselytizing manner, Mancuso's book *Brilliant Green*, published for a general readership, upped the ante by suggesting that not only do plants have the five human senses but they also possess an additional fifteen others, and laid the case for internal and external forms of communication. While acknowledging that plants don't have anything resembling a brain, Mancuso proposed that 'in plants, cognition and bodily functions are not separate but present in every cell,'[7] and that each one is effectively a living internet network, or 'Greenternet', circulating information in a manner such that if one part is destroyed, the rest of the network is still capable of operating. From his perspective, 'they can be described as intelligent. The roots constitute continuously advancing line, with innumerable command centres, so that the whole root system guides the plant like a kind of collective brain – or rather a distributed intelligence – which, as the plant grows and develops, acquires information important to its nutrition and survival.'[8]

Journalist Michael Pollan's long-form article, 'The Intelligent Plant', published in the *New Yorker* in 2013, spread the word to a wider audience, running through the history of research as well as discussing recent developments; he provided another platform for Mancuso's work and drew links with Simard's work on tree communication via underground mycorrhizal networks.

Evolutionary ecologist Monica Gagliano is a passionate proponent of plant intelligence, and keen to make clear that her work is different to that of plant neurobiologists. Fusing science with indigenous perspectives on plants, she is unafraid to put her head above the parapet and use the media to disseminate her take on plant behaviour and findings from empirical research. She describes her book, *Thus Spoke the Plant* (2018), which charts her journey researching plants and learning from shamen, as being written as in collaboration with all of the plants that are part of it.

Gagliano is adamant in her work that the associative learning of plants qualifies them as subjects with cognitive abilities, and that the significance of this demands that 'the current fundamental premise in cognitive science – that we must understand the precise neural underpinning of a given cognitive feature in order to understand the evolution of cognition and behavior – needs to be reimagined.'[9] Dealing with the linguistic conundrum of intelligence, Gagliano draws on the Latin origin of the word to define it as an act of 'choosing between'. Organisms take actions based upon some form of decision making: 'because poor choices are likely to affect performance and survival in many biological systems (including human societies), individuals have evolved a remarkable capacity for making overall good decisions to successfully achieve their ends. This capacity to make sound decisions is not simply hard-wired in a behavioural blueprint, but is a learned skill that can be developed and honed through experience.'[10] She is unambiguous that decision making applies to plants, which participate in an ecology of associative learning: 'When I talk about learning, I mean learning. When I talk about memory, I mean memory.'[11]

Drawing on the work of her predecessors, Gagliano suggests that there could be some type of chemical or hormonal functions at play in the process, but that the result is a distributed form of intelligence that 'leads to one clear, albeit quite different, conclusion: the process of remembering may not require the conventional neural networks and pathways of animals; brains and neurons are just one possible, undeniably sophisticated, solution, but they may not be a necessary requirement for learning.'[12]

Gagliano has also delved into the nascent realm of plant bioacoustics, exploring how plants both respond to sounds and emit them. As her work in the field tentatively reveals:

> In plants, both emission and detection of sound may be adaptive, as preliminary investigations of both processes (in particular reception) suggest. Whilst receptor mechanisms in plants are still to be identified,

> there is early, yet tantalizing, evidence about plants' ability of detecting vibrations and exhibiting a frequency-selective sensitivity that generate behavioural modifications. At both proximate and ultimate levels, sound production in plants is only rarely documented and still poorly understood . . . To date, the production mechanisms and adaptive value of such acoustic emissions remain elusive, yet in the past two decades several studies have pointed to the phenomenological importance of sound and vibrations in plant physiology.[13]

Summarizing the key point of their years of research in a 2021 paper, 'Individuality, self and sociality of vascular plants', Baluška and Mancuso reiterated the importance of vascular systems and root apexes, but added to the arsenal of plant attributes the ability for self-/non-self-recognition, kin recognition and mimicry. Self-/non-self-recognition can be seen in roots and shoots using chemical sensing to discriminate between themselves and other plants in order to aid both plant communication and defence, particularly in competitive environments. Beyond some form of perception of self and adversary, the authors claim that

> socially and cognitively active plants also enjoy kin recognition. Interestingly, plant-specific kin recognition can also be mediated via root exudation. Kin recognition controls new root allocation within root systems in correlation with the distribution and acquisition of nutrients . . . Plants can also recognize kin plants via shoots using photoreceptors. Intriguingly, kin recognition also allows plants to control attraction of pollinators to their flowers. Whether this is accomplished via root exudates or light-sensing photoreceptors is not known . . . As with the root-mediated and exudate-based plant kin recognition, the shoot-mediated and photoreceptor-based plant recognition also plays roles in productivity as those plants which interact with their kin also produce more seeds.[14]

Plant mimicry appears to be something less surprising, given that it has been acknowledged as an attribute for some time, but Baluška and Mancuso suggest that beyond the possibility of chemical sensing, plants are actually likely to use some form of vision to perform the tasks of mimicking shapes, sizes, colours, sizes and textures of different host plants, 'as it is almost impossible to explain mimicking of so many parameters without some kind of vision. It should be not so surprising to have vision-supporting ocelli in vascular plants as eye-like ocelloids are involved in rudimentary vision of unicellular algae.'[15]

As an expression of plant interactions with biotic and abiotic factors, the traits being revealed in this field of research offer the possibility of establishing new metrics for defining plant relationships. If the research advances, and technology provides ways of 'reading' plant signals, this knowledge of plants' learning potential and response adaptations could provide another layer in the understanding of ecological functioning. It may be possible to assume that the composition of plant communities and the arrangements of populations within them in natural habitats have evolved specific spatial distributions to ensure that these modes of communication are optimized to maximize their efficacy and reduce unnecessary drains on plant energy. This could then potentially be factored in at the design stage when considering novel ecosystems and dynamic planting schemes.

Will these responsive abilities aid mitigating change more effectively? Recent research from the old school exponents of the discipline proposes cogent arguments for taking their findings seriously in the face of the climate crisis. Mancuso and Baluška's paper 'Plants, climate and humans' sets out its stall for the need to urgently address our relationship with plants, based upon their active manipulations of other species above and below ground, and the related ecological effects that impact on the climate.[16] The problem may be whether the science is able to keep pace with these types of adaptations. Already due to increased atmospheric pollution, many of the volatile organic compounds are being lost in the particulate soup that surrounds them, meaning that the 'message' they carry to other plants and creatures is getting lost en route. The consequences of warning signals not reaching intended parties or scents alluring pollinators would have a knock-on effect on functioning ecosystems, particularly in cities, where they are most needed. Emissions from vehicles and power plants are known sources of particulate pollution, and levels above a threshold of 80 parts per billion have been identified as problematic.[17] Appreciating the intelligent activities of plants, and trying to understand their forms of communication, will not be easy if their signals are getting lost.

Whatever language is used to describe plant behaviour, the growing body of research shows that the ways in which plants relate to the world are more sophisticated than often assumed. Given the insights into these interactions with stimuli, the question remains, what further revelations will be of mutual benefit in our relationships with plants? Perhaps most significantly, it opens a window into the world of plants through which we can view and appreciate them differently. Accepting that we have more in common with them than previously thought is a gateway to throwing off the fetters of disregard and misconceptions surrounding plants, and to striking new respectful relationships with the environment based on an understanding of their agency.

PLANTS

AS

PARTNERS

TRADITIONAL ECOLOGICAL KNOWLEDGE

Science is reconfiguring the way we comprehend plants and the world we inhabit with them. Far from viewing them as simple immobile organisms, we need to take stock and appraise the strategies that have made them successful survivors across geological time. Having adapted to previous climatic changes, they are doing something similar now, despite the additional challenges we have thrown in their way. The question is, will their strategies be more successful than our own?

The complex interconnections, revealed by research, of plants with their wider environments in ways that affect them both positively and negatively are awe inspiring and eye opening, yet much of this is not new or surprising to many traditional cultures. By its very definition, scientific rationalism is a framework that has denied the validity of knowledge that falls outside of its purview, discrediting the complex weaving of culture and environment that has existed across the globe for millennia. Traditional, local and indigenous forms of knowledge about plants and the rest of nature have shaped the way that societies have lived and responded to the world around them; their ways of understanding parallel some of the findings of science, yet they have arrived at them by completely different routes and challenge many of its preconceptions.

Traditional Ecological Knowledge (TEK) is found in many different forms in disparate cultures and exemplifies the locally rooted connections between people and place. It is a result of close observation and interactions with the environment, in which plants, animals and the land itself are all part of a community. It views each of these as separate – with intrinsic value, sentience and agency – as well as connected to each other and humans. Each act on the other in reciprocal ways, giving and receiving. This connectedness is often viewed as a form of kinship, and they are all accorded respect and honour as relatives and ancestors, and are entwined in cosmologies passed on through oral traditions, rites and practices, binding them together within a shared moral domain.

Understanding and appreciation of plants and their cultural importance is a major part of TEK, as Rarámuri ethnobotanist Enrique Salmon explains: 'The names for plants are markers of our cultural sensitivity to the ecology of the land and to how our culture has embodied that ecological knowledge into our cognitive workings. In addition, cultural references to plant locations, the best sites for plant harvesting, and cultural history add further evidence of how ecological practices have been shaped by the bioregion.'[1] This understanding is embedded in kinship ecology: 'In ceremonial songs kincentricity precipitates from the Rarámuri metaphor that plants are people. The concrete concept of people is mapped onto the more abstract one of plants that can breathe, play, maintain family relationships, and have emotions. With this realization, plants are as important to the Rarámuri as are our nonplant relatives. There is little distinction between our world and that of nonhumans.'[2]

Listening and learning from plants is both a pragmatic and a respectful process that unfolds over time. Drawing on her heritage as a member of the Citizen Potawatomi Nation, environmental biologist Robin Wall Kimmerer describes the process as 'not trying to wring information out of them, but to create situations where they can respond in a way that they can tell the story of their own being and relationship'.[3]

Respecting plants involves harvesting only what is needed from living plants that have the ability to regenerate, and where possible individual plants are not generally destroyed. This prudent use of plants is a caution against depleting populations, and a cross-generational method of living sustainably within the limits of the land. Although different cultures approach the idea in relation to their specific environments, the idea of the 'Honorable Harvest' is a fundamental principle of TEK. Kimmerer describes the practice as 'rules of sorts that govern our taking, shape our relationships with the natural world, and rein in our tendency to consume – that the world might be as rich for the seventh generation as it is for our own'.[4]

While plant science has deliberated over the many possible theories of ecological succession, for many traditional cultures short and long environmental cycles are essential knowledge learnt pragmatically over time and integrated into cultural practices. Creating disturbance by activities such as the regular burning of landscapes in order to regenerate early successional species has become a co-evolutionary bond between people and place.

While these ideas of kinship between humans and the more-than-human world are anathema to the dominant culture of the Western world, they provide alternative perspectives that can inform new relationships. If this is done in a way that avoids cultural appropriation and romanticism, it opens up a space for a range of new engagements with the rest of nature. Already it has been recognized that respecting TEK can be a means of addressing social justice issues, such as the present-day legacy of colonialism and land rights, as well as enhancing environmental protection.

LANDSCAPE LAW

In 1972, Christopher D. Stone, professor of law at the University of Southern California, launched a broadside at the Western idea of human exceptionalism and the notion of personhood in his essay 'Should trees have standing? Toward legal rights for natural objects'. He argued that if an environmental entity is given 'legal personality', it cannot be owned and has the right to appear in court. In Stone's view, 'It is not inevitable, nor is it wise, that natural objects should have no rights to seek redress in their own behalf. It is no answer to say that streams and forests cannot have standing because streams and forests cannot speak. Corporations cannot speak either; nor can states, estates, infants, incompetents, municipalities or universities. Lawyers speak for them, as they customarily do for the ordinary citizen with legal problems.'[1]

While ambitious in his intent, Stone was nonetheless sober in his appreciation of the scope of the task at hand and all it entailed: 'To be able to get away from the view that Nature is a collection of useful senseless objects . . . we have to give up some psychic investment in our sense of separateness and specialness in the universe.'[2] Not only did a sense of separateness need to be abandoned, but also the legal regime of property-based ownership, the cornerstone of capitalism, which affords owners the protection to use, modify and destroy natural features at their own discretion. The same year, a conservation organization, the Sierra Club, took the matter to the US Supreme Court, but to no avail.

Stone's ambition to de-commodify natural features found a resonance with the principle at the heart of TEK, which recognizes and respects the personhood of the rest of nature, and began gaining global momentum. It took a few decades to get a real foothold, but in 2008 Ecuador made a landmark ruling and became the first country to enshrine legal rights to nature in its national constitution. Emboldened by this, local citizens have used it to bring lawsuits against companies causing environmental damage, including a successful case against a construction project on behalf of the Vilcabamba River.

Bolivia followed suit in 2011, spurred on by a resurgence in indigenous Andean religious culture in which the Earth Mother goddess Pachamama is at the centre of all life. The government granted the right of nature 'to not be affected by mega-infrastructure and development projects that affect the balance of ecosystems and the local inhabitant communities', as a way of giving a voice to local communities, and also encouraging new conservation measures to reduce pollution and address climate change issues.[3]

Aotearoa (New Zealand) has been the most progressive nation in addressing these matters. In 2014, the government recognized the legal rights of the Te Urewera forest, changing its status from a national park to its own legal entity, which will own itself in perpetuity with a board to speak as its voice to provide governance and management (Plate 47). The legal act preserves and protects

> Te Urewera for its intrinsic worth, its distinctive natural and cultural values, the integrity of those values, and for its national importance, and in particular to: strengthen and maintain the connection between Tūhoe and Te Urewera; and preserve as far as possible the natural features and beauty of Te Urewera, the integrity of its indigenous ecological systems and biodiversity, and its historical and cultural heritage; and provide for Te Urewera as a place for public use and enjoyment, for recreation, learning, and spiritual reflection, and as an inspiration for all.[4]

It recognizes the great cultural significance of the area for the local Māori Tūhoe community. For similar reasons, the Whanganui River was recognized in 2017 as the legal entity Te Awa Tupua: 'a living and indivisible whole comprising the Whanganui River from the mountains to the sea, incorporating its tributaries and all its physical and metaphysical elements'.[5] Its sense of cultural 'inalienable connection' to the local people is summed up in the proverb 'Ko au te awa, ko te awa ko au' (I am the river and the river is me). The ruling signalled an end to a 140-year-old dispute over the river, finally recognizing its significance as an ancestor to the Whanganui community, with corresponding rights, powers, duties and liabilities as a legal person. Two appointed guardians, one each from the tribe and the government, are charged to act on behalf of the river to ensure that infrastructure and development projects will not affect the balance of ecosystems and local communities. The same year, Mount Taranaki became the third landscape entity in the country to be accorded a legal personality (Plate 48).

Further notable legal progress has been made in India, which in 2017 gave the Ganges and Yamuna rivers all the constitutional and statutory rights of human beings, including the right to life, to protect them or keep them clean. A 2019 ruling in Bangladesh by the Supreme Court means to protect all the rivers that form the world's largest delta from further degradation from pollution. The government-appointed National River Conservation Commission has the power to take to court anyone accused of harming it. Part of the Amazon rainforest in Columbia obtained its own rights in 2018, following the granting of rights in 2017 to the Rio Atrato, a river in the northwestern area of the country. In the United States, the state of Ohio in 2019 gave citizens the ability to sue on behalf of Lake Erie whenever its right to flourish is being contravened by environmental harm. In Canada in 2021, joint resolutions by the local municipality of Minganie and the Innu Council of Ekuanitshit granted legal personhood to the Muteshekau-shipu (Magpie River).

Steps to treat plants in the same manner have also been under consideration. In Switzerland in 2008, a report by the government-appointed Federal Ethics Committee on Non-Human Biotechnology (ECNH) described interfering with plants without a valid reason as 'morally inadmissible'. In the United States in 2019, the White Earth Band of Ojibwe and the 1855 Treaty Authority adopted Rights of Manoomin for on- and off-reservation protection of wild rice, and the clean fresh water resources and habitats in which it thrives, in northwestern Minnesota. The dictate identifies Manoomin's rights to exist, flourish, regenerate and evolve, as well as inherent rights to restoration, recovery and preservation, in a natural environment free from industrial pollution, human-caused climate change impacts, patenting, and contamination by genetically engineered organisms (Plate 49).

The granting of the status of personhood addresses a number of important social and environmental concerns. As a conservationist policy it aims to prevent landscape degradation, while culturally it recognizes marginalized communities and the legacy of colonialism. The promotion to personhood of more-than-human entities can be read as the latest stage in Western liberalism's progressive redressing of imbalances, following in the footsteps of addressing rights around ethnicity, gender, sexuality and some other species. Some critics have raised concerns that while it has positive effects, it domesticates parts of the natural world within an anthropocentric legal and political framework based on their instrumental and cultural values, but ultimately still denies their intrinsic characteristics and qualities.

But opening the debate and putting rights into a wider context highlights a significant broaching of the gulf between Western and indigenous ways of thinking, which have previously been incompatible. Levelling the playing field between humans and more-than-human beings challenges the unified 'otherness' traditionally ascribed to the rest of nature and the separation of people from it. This legal sea change runs parallel to a paradigm shift that has been taking place within the academic disciplines of the social sciences and humanities in the Global North in recent years.

THE ONTOLOGICAL TURN

The move towards recognizing and respecting TEK by using legal systems can be seen as a wider move towards social and environmental equity, which has its roots in local struggles. But acknowledgement of these wider issues has also been aided by what has been termed the 'ontological turn', which has been taking place in the humanities and social sciences since the 1990s. This disruptive intervention has sought to dismantle binary thinking and hierarchies by viewing reality in a more relational way, recognizing that humans' relationships with the world around them are more complex and interrelated. In particular, this turn has provided a serious challenge to the splits established by Enlightenment rationalism between nature and culture, human and more-than-human, and subject and object, thereby calling for a rethinking of our place in the world and relationships with everything else within it.

In the field of anthropology, the methodological perspectives developed in France by Phillipe Descola, Bruno Latour and Viveiros de Castro spurned the traditional approach of the discipline in which the activities of indigenous cultures were seen as analogues for Western ones, and the job of the anthropologist was simply to translate them to make sense of the culture. Traditional epistemologies and cosmologies were viewed as metaphors for actual things in the world, which were comprehensible when translated across cultures; this view reflected an unconscious bias in which it was assumed that all cultures thought and behaved according to Western ways of doing things. Instead, this new approach rejected such a universalizing way of thinking in favour of the multiple ways of knowing the world of sensory experience, which are inseparable from being in it. Consequently, myths and stories were no longer viewed as metaphors for something else, but as unique understandings of the world imbued with their own sense of integrity. Rather than being a mistaken belief in the workings of the natural world, the idea of animism became understood instead as a form of ontological engagement with the world, a way of being with its own validity outside the strictures of Western thinking.

Drawing on traditional knowledge in ethnographic studies in the Amazon, Descola and de Castro's work reconfigured the idea of the environment as a passive background into a living pluriverse of agency, in which humans are but one type of actor. Descola introduced a rupture into the nature/culture spilt by drawing on the traditional views of the Achuar culture, in which the distinction didn't exist. His 2005 book, *Beyond Nature and Culture*, dissolved the category of nature, and proposed a new framework for understanding the world featuring 'four ontologies': animism, totemism, naturalism and analogism. Describing the outlook of a number of different societies, de Castro proposed a form of perspectivism, using the term 'multinaturalist' to describe the way that all beings, human and more-than-human, share the same culture, soul or perspective, but differ across the bodies they possess and the worlds that they perceive.

Latour's Actor Network Theory sought to undermine the established basis of rationality premised upon the split between the mind and body by according all matter with agency, in an equalizing flat ontology. In *We Have Never Been Modern* (1991), his critique of contemporary rationality suggested that 'modern humanists are reductionist because they seek to attribute action to a small number of powers, leaving the rest of the world with nothing but simple mute forces.'[1] Stressing the importance of the effects of such a flattening beyond simply an anthropocentric outlook, anthropologist Tim Ingold reflected:

> In a more-than-human world, nothing exists in isolation. Humans may share this world with non-humans, but by the same token, stones share it with non-stones, trees with non-trees and mountains with non-mountains. Yet where the stone ends and its contrary begins cannot be ascertained with any finality . . . My attention is rather directed towards a place from which I see something happening, a going-on that spills out into its surroundings, including myself. We should replace our nouns for naming things with verbs: 'to stone', to tree', 'to mountain', to 'human'. At once the world we inhabit, and that we share with so many other things, no longer appears ready-cut, into things of this sort or that, along the lines of classification.[2]

The active nature of everything in the world is at the heart of anthropologist and biologist Donna Haraway's early work on living with other companion species, which set the stage for a focus on everyday relationships with species including 'such organisms as rice, bees, tulips, and intestinal flora, all of whom make life for humans what it is – and vice versa'.[3] Haraway believes reality is an active verb in which we are situated in entanglements with the multitude of more-than-human 'critters' around us. This necessitates an ongoing process of 'worlding' that is an active and embodied engagement of giving presence to other species in our relationships with them in acts of making kin, denying simple resolutions, and stretching the boundaries between fact and fabulation. Haraway outlined the complexity of these entanglements in her 2016 book, *Staying with the Trouble*.

Further developments in anthropology have taken place under the moniker of Multispecies Ethnography, exploring the interconnectedness of humans and other life forms, including plants, animals, fungi, bacteria and viruses, and showcasing the intersections of the relationships between ecology, politics, economics and cultural practices.

Australia-based ethnographer Deborah Bird Rose's work on using storytelling as a means to address issues of extinction draws on her research with the Aboriginal peoples of the Victoria River region of the Northern Territory. Telling the stories of threatened species has the potential to draw others into new relationships and accountabilities with them; it offers a means to develop 'lively ethographies', or ways of knowing, engaging and appreciating the lives of other species by providing a framework of 'attentiveness to the evolving ways of life (or ēthea; singular: ethos) of diverse forms of human and nonhuman life . . . in an effort to explore and perhaps restore the relationships that constitute and nourish them'.[4]

Anthropologist Anna Tsing's work delves into the landscapes of the Anthopcene shaped by the practices and legacies of resource extraction industries. Her investigations in *The Mushroom at the End of the World* (2015) led her to unravel a complex global network of inseparable ecological and social relationships. Drawing connections between environmental disturbances as a product of capitalist environmental degradation and the resulting co-dependency of multispecies assemblages, she weaves together labour relations with the symbiotic associations of fungi and tree roots to describe novel ecosystems featuring both human and more-than-human participants. Reflective of the all-encompassing commodification of natural resources in the Anthropocene, these ecosystems are novel not by design but by consequence, 'because human disturbance makes the presence of matsutake more likely – despite the fact that humans are entirely incapable of cultivating the mushroom. Indeed, one could say that pines, matsutake, and humans all cultivate each other unintentionally. They make each other's world-making projects possible.'[5] The idea that these landscapes are co-created breaks down human hubris, placing people within, rather than outside, natural processes in a world of overlapping agency.

CRITICAL PLANT STUDIES

This ontological perspective, which flattens hierarchies, has ramifications for plants. The level playing field that it creates lines humans up alongside them rather than above them in a superior manner. It also reveals that plants' existence is as equally meaningful for other species as it is for ours, suggesting an ethical problem within our relationships with them. These ideas have been pushed further in the field of Critical Plant Studies, emerging as a cross-disciplinary area, challenging a lack of plant awareness on every level, and expanding the notions of subjectivity and agency.

Philosopher Michael Marder has been a prolific critic of anthropocentric outlooks, advocating the need to think vegetatively. By deconstructing the history of philosophy, his work has challenged the exclusion of plants through ascribing a state of otherness to them, and the opposition between 'logocentric' and 'biocentric' approaches to the world. He explores germination, growth, blossoming, fruition, reproduction and decay as illustrations of abstract concepts, along with ideas from plant biology and studies in plant sentience; his work aims to rethink cultural and instrumental attitudes in favour of a more respectful approach to vegetal beings, examining 'the possibility of an ethical treatment of plants grounded in empathy'.[1] This is particularly important in the current global environmental crisis, when massive deforestation, seed patenting and profit-driven agriculture threaten the future of many forms of life on the planet. Marder, collaborating with French philosopher Luce Irigaray, looks towards plant characteristics as expressions of alternative ways of thinking about the world: 'Take, for example, the so called "modal development" of plants that grow by branching out in every conceivable direction. Such open-ended growth reveals that plants are neither machines nor organisms, subordinated to the whole external to them or to a pre-existing plan. Our human growth could perhaps take them as a model in order not to merge into a whole, in which each might vanish (almost) without a remainder.'[2] He views this critical perspective as a manifesto essential for formulating action: 'The plant is immanently de-centered, and so are the philosophy and the ethics that put it in the limelight. A community of growth within wider communities of growth, it should provide us with a concrete model for political organization and cohabitation, thought and action. That is the promise of phytocentrism to come.'[3]

Unlike Marder's promotion of the uniqueness of the vegetable world, botanist Matthew Hall, in *Plants as Persons* (2011), has instead traced the trajectory of plants through the history of ideas and global religions in a bid to resituate them in their relationship to people. Arguing on their behalf, Hall highlights the fact that, given a driving life force is a common denominator, 'plants and humans share a basic, ontological reality as perceptive, aware, autonomous, self-governed, and intelligent beings. Like other living beings, plants actively live and seek to flourish. They are self-organized and self-created as a result of interactions with their environment.[4] Plant science confirms the ways in which plants sense and respond to their environments, to one another and to other species in symbiotic ways, yet their autonomy is ignored, and the ongoing deliberate disrespect they are subjected to flies in the face of adopting the new types of environmental ethics needed for the future. Such ethics need to be able to include 'plants within the realm of moral consideration; for the sake of individual plants and plant species and for those animals and humans whose lives depend on their survival'.[5]

Thinking about plants in more specific contexts, geographers Owain Jones and Paul Cloke have looked towards trees and the ways that they simultaneously shape experiences, ascribing to them a sense of agency often unrecognized by the people who experience their effects. They suggest trees' ability

> to grow, reproduce, spread, break up monuments, figure significantly in the emotions of nearby residents, demarcate heritage and so on have in each case slipped the leash of human plans. The trees have acted as relatively autonomous material presences which have spanned across and between eras of place identity and place configuration. In doing so, their powerful material presence has relationally shaped the new place identities and configurations that have emerged. New waves of politics, emotions, economics and governance have gathered around the trees and formed alliances (or otherwise) with them in disputes about future place form.[6]

For Jones and Cloke, trees are critical players in the dynamic landscape, mediating and forging alliances that define local histories: 'Our contention is that nature–society relations are continually unfolding in the contexts of specific places, in which meanings will arise from particular interactions between different assemblages of social, cultural and natural elements.'[7]

Canadian geographer Renate Sander-Regier extends the notion of agency to plants in general and locates the garden as an important nexus of interspecies activity, in which 'the gardener is well positioned to experience an accompanying diversity of plant agency on a continuous and intimate basis – often admiring and celebrating it, frequently also struggling with it.'[8] The unique and sometimes subtle actions of plants are played out through rhythms and timescales specific to botanical being, reflecting their distinct vitality, clear to those gardeners who take time to stop and notice. Acknowledging the garden as a hybrid space of co-evolution where plants have agency over their own lives and influence over those of humans also, she suggests that they provide opportunities to know more about 'these botanical partners in the garden project – we must, if only to satisfy our own curiosities about the non-human Others that live among us, to acknowledge and accept "the livingness of the world", and to start thinking differently about the other life forms that share the planet.'[9]

ETHICS IN THE ANTHROPOCENE

The current confluence of knowledge from different disciplines and practices has created the perfect storm for a reappraisal of plant life, in terms of both the lives that plants actually lead and our lives with plants. And it couldn't have come at a more opportune moment. The sum on the balance sheet makes it clear that the deficit of our relationships with plants needs immediate address. The opportunities are there, but it's more than a matter of just acknowledging underappreciated aspects of plant lives; we have to want to effect change on all levels and at all scales, and to care enough to actually do it.

The latter is too often the sticking point. The ways in which we use self-consciousness often work against us, as is evident from the situation we currently find ourselves in. It creates self-interest, and our individual concerns are so often in conflict with those of others, human and more-than-human. The Global North has reinforced a system of thinking over the past five hundred years in which self-interest is the default position and something to be admired. Individualism undermines so many of our communal aspirations, the ones from which we all can benefit, and which we need to realize in order to survive. Individualism is the mindset of the Anthropocene, the take-it-all-until-it's-gone, smash and grab short-sightedness, that places personal profit above all else.

This exemplifies one end of the spectrum of ethical views of the world. Egocentrism is the dominant Western capitalist mode of thinking, in which human exceptionalism confers privilege on *Homo sapiens* as the most important species on the planet. The extractive mindset this harbours views the rest of nature as resources to be plundered at will, and has led to the climate and biodiversity crises, whether we subscribe to the outlook or not. At the other end is Ecocentrism, which denies divisions between humans and the rest of nature and ascribes intrinsic value equally. As a preservationist perspective, it places the rest of nature above human considerations and views humans as a threat to be kept at a distance. It is closely associated with the idea of Deep Ecology, initiated by Norwegian philosopher Arne Næss. In between these two poles, Biocentrism places inherent value on other species, although these are often hierarchically ranked and are still considered in relation to human needs. Management and sustainability are key to this position.

But the notion of nature within each of these perspectives is problematic as it still invokes the binary separation between humans and the rest of nature rather than incorporating them within an integrated outlook. We have old and new ways of thinking about our place in the world to draw upon in rethinking our relationships with it. Alternative ontologies and epistemologies present us with expanded forms of thinking, from which we can shape new forward-facing ethical outlooks.

What is required is an embodied ethics in our daily encounters and long-term relationships with plants. We can appreciate them for their sensual attributes and as sophisticated ecological engineers, but we need to stop treating them simply as resources and commodities, eradicating them out of habit, and harrying them into arrangements simply for our own delectation. Plants are more than purchases to push up property prices. We need to stop arranging them like furniture to match the wallpaper of our consumer lifestyles.

Drawing on our co-evolutionary bond with plants, gardeners already have intimate associations with them ranging from casual interest to deeply engaged and emotional interactions. They already have their sleeves rolled up and hands dirty, and are well placed to consider our ongoing bonds.

A PLANT-CENTRIC PERSPECTIVE: SOWING SEEDS FOR THE FUTURE

Our knowledge of ourselves and the rest of the world is changing at a rapid pace. We are gaining increasing insight into the intricacies and complexities of ecological relationships, the interconnections between species, and the abiotic environment. We are learning more about their dynamic and ever-changing nature, providing us with greater understanding of the subtlety and unpredictability of so many of these relationships. We need to realize that accepting both anticipated outcomes and random effects are part of the process of life itself. But will we be able to do this at a pace fast enough to outrun the larger-scale change we have effected through our own actions due to our discordant relationship with the rest of nature? Will we just keep ignoring the everyday extinctions going on all around us as we go about business as usual?

In order to face the challenges we are confronted with by the climate and extinction crises, we need to ask ourselves some serious questions. It is obvious that we cannot simply continue with the same anthropocentric arrogance, and that we need to reconsider our place on the planet and those of everything else around us. What will this mean for the personal habitats we create and those we share them with? How will we garden our personal, public and planetary plots? Plants are at the heart of the world we have evolved in, and as primary producers shape the way we live in it. They have successfully adapted and survived dramatic changes throughout the long duration of their presence on the planet. They are the ever-present allies we simply take for granted or ignore. We need to adopt and embrace a plant-centric perspective which informs everything we do, and place plants centre stage, accorded the respect they are rightly due, rather than treat them merely as resources and commodities in instrumental or aesthetic ways.

It's time to put down roots with our neighbours. We should not simply ask what more they can do for us, but also ask what we can do for them in order to work as partners. In the Anthropocene being seen to be green – by just paying lip service to plants – is not enough, yet it has become part of everyday political parlance, which we need to get beyond.

Understanding plants and engaging with them in new ways is a step towards living in more ecologically astute ways, embracing the dynamic nature of the world. It means leaving behind an anthropocentric view and accepting that total control of the world around us is an unobtainable illusion. Part of the problem of *Homo sapiens* is that we think we can know everything – it's inbuilt into our drive for dominance and leads us to make unrealistic assumptions. Technology amplifies this and assures us that the future is knowable, that artificial intelligence will assist us in our lust for total knowledge. Yet as ecology reveals, the more we discover, the more we realize that there is still more to be found.

How we shape our spaces, and the ways we live with plants in domestic gardens, public parks and semi-natural landscapes, play a major role in how ecosystems function. We no longer need to choose between an aesthetic approach and an ecological one; we have a palette of possibilities that offer opportunities for both approaches in traditional and novel situations. What would the world look like if the criteria for selecting and arranging plants were to be based not simply on the colour of flowers, but also on the insights of plant science, considerations of mutualistic microbes and the smart actions of plants themselves? That would depend on the skill and artistry applied to utilizing ecological knowledge. It may not be too visually different to now, but the differences in how it functions and the benefits it brings could be significant. As contemporary practitioners show, a paradigm shift is underway and the momentum it is building can put plants at the forefront of ensuring that the ecosystems around us can function in all the different ways that they do. Gardeners have absorbed horticultural homilies by rote and carried them out as though they were second nature. The next step is to do the same with ecological knowledge, putting it into everyday practice, aligning the interests of people and plants.

We all have responsibilities as gardeners in the widest sense of the word, and our actions matter. Working more wisely with plants will prepare us for confronting the challenges of the future. It should be a cause for celebration not a chore. We have a plethora of possibilities afforded by many types of knowledge, and digging deeper will reveal even more. The changing climate enfolds us in the vastness of its vicissitudes, placing our purported prowess in a bigger perspective. When we use our unique species trait, the ability to look into the future, how do we see plants there beside us?

NOTES

INTRODUCTION

1 Anthropocene Working Group, 'Subcommission on Quaternary Stratigraphy', 2019, www.quaternary.stratigraphy.org.
2 Jason W. Moore, 'The Capitalocene, part I: On the nature and origins of our ecological crisis', *Journal of Peasant Studies*, vol. 44, no. 3 (2017), p. 594, doi.org/10.1080/03066150.2016.1235036.
3 'Effects of global warming', *National Geographic*, 14 January 2019, www.nationalgeographic.com/environment/article/global-warming-effects.
4 Lori Cuthbert, 'The extraordinary ways weather has changed human history', *National Geographic*, 30 May 2018, www.nationalgeographic.com.
5 Scripps Institution of Oceanography, The Keeling Curve, 2K Years, 'Ice core data before 1958, Mauna Loa data after 1958', https://keelingcurve.ucsd.edu (accessed 13 February 2022).
6 NOAA, Earth System Research Laboratories (ESRL), 'Trends in atmospheric carbon dioxide', https://gml.noaa.gov/ccgg/trends; 'Trends in atmospheric methane' https://gml.noaa.gov/ccgg/trends_ch4; 'Trends in atmospheric nitrous oxide' https://gml.noaa.gov/ccgg/trends_n2o (accessed 13 February 2022).
7 'Working Group I: The scientific basis', International Panel on Climate Change (IPCC) https://archive.ipcc.ch/ipccreports/tar/wg1/016.htm.
8 Michael Le Page, 'CO_2 set to hit levels not seen in 50 million years by 2050', *New Scientist*, 4 April 2017, www.newscientist.com/article/2126776-co2-set-to-hit-levels-not-seen-in-50-million-years-by-2050.
9 Robert McSweeney, 'Scientists solve ocean "carbon sink" puzzle', *Carbon Brief*, 8 February 2017, www.carbonbrief.org.
10 Elizabeth Kolbert, *The Sixth Extinction*, London, Bloomsbury, 2014, p. 114.
11 UNESCO, 'Assessment: World Heritage coral reefs likely to disappear by 2100 unless CO2 emissions drastically reduce', 23 June 2017, whc.unesco.org/en/news/1676.
12 NOAA, National Centers for Environmental Information, Global Climate Report – Annual 2020, www.ncdc.noaa.gov/sotc/global/202013.
13 IUCN, 'The IUCN Red List of threatened species', version 2021–2, 9 December 2021, www.iucnredlist.org.

PLANTS AS PRODUCERS

IN PRAISE OF PLANTS

1 Andrea Thompson, 'Plants are the world's dominant life–form', *Scientific American*, 1 August 2018, www.scientificamerican.com/article/plants-are-the-worlds-dominant-life-form.

PLANTS AND A CHANGING PLANET

1 S. Piao et al., 'Plant phenology and global climate change: Current progresses and challenges', *Global Change Biology*, 18 March 2019, doi.org/10.1111/gcb.14619.
2 U. Büntgen et al., 'Plants in the UK flower a month earlier under recent warming'. *Proceedings of the Royal Society B*, vol. 289, no. 1968 (2022), doi.org/10.1098/rspb.2021.2456.
3 Amelia A. Wolf, Erika S. Zavaleta and Paul C. Selmants, 'Flowering phenology shifts in response to biodiversity loss', *PNAS*, vol. 114, no. 3 (28 March 2017), pp. 3463–8, first published 13 March 2017, doi.org/10.1073/pnas.1608357114.

PLANTS AS PANACEA

THE UNSEEN GREEN

1 J. H. Wandersee and E. E. Schussler, Preventing plant blindness', *American Biology Teacher*, vol. 61, no. 2 (February 1999), p. 82.
2 C. McDonough MacKenzie et al., 'We do not want to "cure plant blindness", we want to grow plant love', *Plants, People, Planet*, vol. 1, no. 3 (2019), pp. 139–41.
3 Ricardo Rozzi, 'Biocultural ethics: Recovering the vital links between the inhabitants, their habits, and habitats', *Environmental Ethics*, vol. 34, no. 1 (2012), pp. 27–50. Juan L. Celis-Diez et al., 'Biocultural homogenization in urban settings: Public knowledge of birds in city parks of Santiago, Chile', *Sustainability*, vol. 9, no. 4 (2017), p. 485.
4 Robert Macfarlane and Jackie Morris, *The Lost Words: A Spell Book*, London, Hamish Hamilton, 2017.

HEALTH AND WELL . . . BEING

1 Jake Robinson, Jacob Mills and Martin Reed, 'Walking ecosystems in microbiome-inspired green infrastructure: An ecological perspective on enhancing personal and planetary health', *Challenges*, vol. 9, no. 2 (2018), p. 40.
2 Cecily Maller et al., 'Healthy nature healthy people: "Contact with nature" as an upstream health promotion intervention for populations', *Health Promotion International*, vol. 21, no. 1, (2005), pp. 45–54.
3 Tony McMichael, *Human Frontiers, Environments and Disease: Past Patterns, Uncertain Futures,* Cambridge University Press, 2001, p. 232.
4 Ibid., p. 2.
5 Glenn Albrecht, '"Solastalgia": A New Concept in Health and Identity', *PAN: Philosophy, Activism, Nature*, no. 3 (2005), p. 48.

THE NATURE DISCONNECT

1 Raymond Williams, *Keywords,* Oxford University Press, 1976, p. 184.
2 Richard D. Ryder, 'Speciesism again: The original leaflet', *Critical Society,* no. 2 (Spring 2010), p. 1.
3 John Locke, *The Second Treatise on Civil Government* (1689), ed. Thomas Preston Peardon, New York, The Liberal Arts Press Inc., 1952, p. 17.
4 Carolyn Merchant, *The Death of Nature: Women, Ecology and the Scientific Revolution,* New York, Harper & Row, 1980, p. 2.
5 Val Plumwood, *Feminism and the Mastery of Nature*, New York, Routledge, 1993, p. 42.
6 Val Plumwood, 'Nature, self and gender: Feminism, environmental philosophy and the critique of rationalism', in *Environmental Ethics: An Introduction with Readings*, ed. John Benson, New York, Routledge, 2000, p. 264.
7 Deborah Bird Rose, 'Nature as power', University of New South Wales, 2016, www.futurelearn.com/info/courses/remaking–nature/0/steps/16724.
8 Lynn Margulis and Dorion Sagan, *Microcosmos*, New York, Summit Books, 1986, p. 22.
9 J. Baird Callicott, 'La Nature est morte, vive la nature!', *Hastings Center Report*, vol. 22, no. 5 (1992), p. 18.
10 J. M. Robinson, J. G. Mills and M. F. Breed, 'Walking ecosystems in microbiome-inspired green infrastructure: An ecological perspective on enhancing personal and planetary health, *Challenges*, vol. 9, no. 2 (2018), p. 40, doi.org/10.3390/challe9020040.

MANAGING THE ENVIRONMENT

1 Herman E. Daly, 'On economics as a life science', *Journal of Political Economy*, vol. 76, no. 3 (1968), p. 397.
2 Donella H. Meadows et al., *The Limits to Growth*, New York, Universe Books, 1972, p. 158.
3 E. F. Schumacher, *Small is Beautiful: Economics as if People Mattered*, London, Blond & Briggs, 1973/Perennial Library, 1975, p. 51.
4 Gro Harlem Brundtland, *Our Common Future*, Geneva, World Commission on Environment and Development, 1984, p. 11.
5 Ibid., p. 41.

ECOSYSTEM SERVICES

1 Gretchen Daily, 'Introduction: What are ecosystem services?', in *Nature's Services: Societal Dependence on Natural Ecosystems*, ed. Gretchen Daily, Washington, D.C., Island Press, 1997, p. 3.
2 Robert Constanza et al., 'The value of the world's ecosystem services and natural capital', *Nature*, vol. 387 (15 May 1997), p. 253.
3 European Commission, 'Nature-based solutions', https://ec.europa.eu/info/research-and-innovation/research-area/environment/nature-based-solutions_en.
4 Ibid.
5 L. A. Roman et al., 'Beyond "trees are good": Disservices, management costs, and tradeoffs in urban forestry', *Ambio*, vol. 50 (2021), pp. 615–30, doi.org/10.1007/s13280-020-01396-8.

ENVIRONMENTAL PRACTICE

1 Half-Earth Project, 'Why half?', www.half-earthproject.org/discover-half-earth.
2 Michael Soulé and Reed Noss, 'Rewilding Biodiversity', *Wild Earth*, vol. 8, no. 3 (1998), p. 19.
3 IUCN, CEM Rewilding Thematic Group, 'Rewilding principles', https://www.iucn.org/commissions/commission-ecosystem-management/our-work/cems-thematic-groups/rewilding (accessed 13 February 2022).

NOVEL ECOSYSTEMS

1 Stephen Jay Gould, 'An evolutionary perspective on strengths, fallacies, and confusions in the concept of native plants', in *Nature and Ideology, Natural Garden Design in the Twentieth Century,* ed. Joachim Wolschke-Bulmahn, Washington, D.C., Dumbarton Oaks Research Library and Collection, 1997, p. 15.

2 Scott P. Carroll, 'Conciliation biology: The eco-evolutionary management of permanently invaded biotic systems', *Evolutionary Applications*, vol. 4, no. 2 (2011), p. 184, doi:10.1111/j.1752-4571.2010.00180.x.
3 Ibid., p. 187.
4 Michael L. Rosenzweig, 'Reconciliation ecology and the future of species diversity', *Oryx*, vol. 37, no. 2 (April 2003), p. 201.

URBAN ECOLOGY

1 Richard Mabey, *The Unofficial Countryside*, London, William Collins & Sons, 1973.
2 Herbert Sukopp, 'On the early history of urban ecology in Europe', *Prelia*, vol. 74 (2002), p. 381.
3 Herbert Sukopp, 'Urban ecology: Scientific and practical aspects', in *Urban Ecology*, ed. J. Breuste, H. Feldmann and O. Uhlmann, Berlin, Springer-Verlag, 1998, p. 5.

GARDEN ECOLOGY

1 Ministry of Housing, Communities & Local Government, 'Land Use in England, 2018', Planning Official Statistics Release, 16 July 2020, p. 4.
2 Ministry of Housing, Communities & Local Government, 'English Housing Survey, Headline Report, 2019–20', December 2020, p. 3.
3 Jennifer Owen, *The Ecology of a Garden: The First Fifteen Years*, Cambridge University Press, 1991, p. 345.
4 Ibid.
5 Ken Thompson et al., 'Urban domestic gardens (I): Putting small-scale plant diversity in context', *Journal of Vegetation Science*, vol. 14 (2003), pp. 71–8.
6 James Hitchmough and Nigel Dunnett, 'Introduction to naturalistic planting in urban landscapes', in *The Dynamic Landscape: Design, Ecology and Management of Naturalistic Urban Planting*, ed. Nigel Dunnett and James Hitchmough, London and New York, Spon Press, 2004, p. 13.

CLEANING UP THE GARDEN

1 In 1943, the Germans, while experimenting with military gases during the war, made a synthesis of the first organophosphate insecticide called parathion.
2 Rachel Carson, *Silent Spring*, London, Penguin, 1962.
3 Dave Goulson, *The Garden Jungle*, London, Jonathan Cape, 2019, p. 75.
4 DEFRA, 'Ending the retail sale of peat in horticulture in England and Wales', December 2021, www.consult.defra.gov.uk.
5 Ibid.
6 Royal Horticultural Society, 'RHS peat policy', 2021, www.rhs.org.uk/about-the-rhs/policies/rhs-statement-on-peat.
7 David Bek, 'The future for plastics in the retail supply chain', Plastics in Ornamental Horticulture: Creating a Sustainable Supply Chain, AIPH Sustainabilty Conference 2019, www.aiph.org.
8 Marie Soulliere-Chieppo, 'Plastic pots and the green industry' Association of Professional Landscape Designers, 2020, www.cdn.ymaws.com.

PLANTS AS PICTURES

GROWING THE IDEA OF THE GARDEN

1 John Dixon Hunt, *Greater Perfections*, Cambridge, Mass., MIT Press, 2000, p. 62.
2 Derek Clifford, *A History of Garden Design*, London, Faber & Faber, 1962, p. 15.
3 Richard Mabey, *Weeds*, London, Profile Books, 2010, p. 1.
4 Harriet Ritvo, 'At the edge of the garden: Nature and domestication in eighteenth and nineteenth-century Britain', *Huntington Library Quarterly*, vol. 55, no. 3 (1992), p. 367.
5 John Passmore, *Man's Responsibility for Nature: Ecological Problems and Western Traditions*, New York, Scribner, 1974, p. 36.
6 Isabelle Van Groeningen, 'The Development of Herbaceous Planting in Britain and Germany from the Nineteenth to Early Twentieth Century', PhD thesis, University of York, 1996.
7 William Cobbett, *The English Gardener*, London, W. Cobbett, 1829, Chapter VII, para. 411.
8 Alexander Pope, Epistle IV, 'To Richard Boyle, Earl of Burlington, of the uses of riches' (1731), in *The Works of Alexander Pope Esq., Volume 3, containing his Moral Essays*, London, J. and P. Knapton, 1751, p. 186.
9 Ibid., p. 187.
10 Horace Walpole, 'On modern gardening', in *Anecdotes of Painting in England*, vol. 4, London, Strawberry Hill, 1771, p. 138.
11 Thomas Jefferson, 'Notes of a tour of English gardens, [2–14 April] 1786', https://digitalarchive.wm.edu/handle/10288/15264.

12 Humphry Repton, *An Enquiry into the Taste in Landscape Gardening*, London, J. Taylor, 1806, p. 215.
13 Ibid.
14 Ibid., p. 23.
15 Humphrey Repton, *The Landscape Gardening and Landscape Architecture of the Late Humphrey Repton, Esq.*, ed. J. C. Loudon, London and Edinburgh, Longman & Co. and A. & C. Black, 1840, p. 171.
16 J. C. Loudon, Introduction to Repton, *Landscape Gardening and Landscape Architecture*, p. viii.
17 J. C. Loudon, 'On laying out and planting the lawn, shrubbery, and flower-garden', *Gardeners Magazine*, vol. IX (1843), p. 167.
18 William Robinson, 'The London parks: No. I, Battersea Park', *Gardeners' Chronicle and Agricultural Gazette*, September 1864, pp. 843–4.
19 William Robinson, *Garden Design and Architects' Gardens*, London, John Murray, 1892, p. 22.
20 William Robinson, 'Bedding out, a defence and a reply, part I', *The Garden*, vol. II (September 1872), pp. 264–5.
21 William Robinson, *The Wild Garden*, 3rd edn, London, John Murray, and New York, Scribner and Welford, 1883, p. 7 (first published 1870).
22 Ibid., p. 163.
23 Ibid., p. 33.
24 Robinson, *Garden Design and Architects' Gardens*, p. 9.
25 Robinson, 'Bedding out', pp. 264–5.

THE COLOURISTS

1 Gertrude Jekyll, *Home and Garden*, London, Longmans, Green, and Co., 1900, p. 45.
2 Gertrude Jekyll, *Colour in the Flower Garden*, London, Country Life/George Newby, 1908, p. vii.
3 Gertrude Jekyll, *Wood and Garden: Notes and Thoughts, Practical and Critical, of a Working Amateur,* London, Longmans, Green, and Co., 1899, p. 264.
4 Ibid., p. vi.
5 Louise Beebe Wilder, *Colour in My Garden*, New York, Doubleday, Page & Company, 1918, p. 4.
6 Fletcher Steele, 'Color charts', *Bulletin of the Garden Club of America*, no. VIII (March 1921), p. 17.
7 Thomas Mawson, *The Art and Craft of Garden Making*, London, B. T. Batsford and George Newnes, 1900, p. 15.
8 Vita Sackvile-West, *Let Us Now Praise Famous Gardens*, London, Penguin, 2009, p. 48.
9 Vita Sackville-West, *V. Sackville West's Garden Book*, London, Michael Joseph, 1968, p. 15.
10 Ibid., p. 91.
11 Vita Sackville-West, *The Illustrated Garden Book*, London, Michael Joesph, 1986, p. 149.
12 Margery Fish, *Cottage Garden Flowers*, London, Faber & Faber, 1980, p. 9 (first published 1961).
13 Margery Fish, *An All the Year Garden*, Newton Abbott, David & Charles, 1972, p. 9 (first published 1958).
14 Rosemary Verey, *The English Country Garden*, London, BBC Books, 1996, p. 40.
15 Penelope Hobhouse, *Colour in Your Garden*, New York, Little, Brown and Company, 1985, p. 18.
16 Elsie Swerhorn, 'Recreating Eden: Episode 1 – The Colour of Love (Sandra and Nori Pope)', 2002, YouTube 2020, www.youtube.com/watch?v=ATcTWOsc8DE.
17 Ibid.
18 Nori and Sandra Pope, *Colour by Design*, London, Conran Octopus, 1998, p. 11.

MODERNISM: FORM, FUNCTION, FOLIAGE

1 James C. Rose, 'Plants dictate garden forms', *Pencil Points*, vol. 19 (November 1938), p. 695.
2 Ibid., p. 697.
3 Ibid.
4 Ibid., p. 695.
5 Garrett Eckbo quoted in Mark Francis and Randolph T. Hetser, Jr, *The Meaning of Gardens*, Cambridge, Mass., MIT Press, 1990, p. 56.
6 Garrett Eckbo, *Landscape for Living*, Santa Monica, Calif., Hennessey & Ingalls, 2002, p. 95 (first published 1950).
7 Garrett Eckbo, *The Art of Home Landscaping*, New York, F. W. Dodge, 1956, p. 236.
8 Eckbo, *Landscape for Living*, p. 95.
9 Ibid.
10 Ibid., p. 96.
11 Thomas D. Church, *Your Private World: A Study of Intimate Gardens*, San Francisco, Chronicle Books, 1969, p. 30.

12 Christopher Tunnard, *Gardens in the Modern Landscape*, London, Architectural Press, 2nd edn, 1948, p. 126 (first published 1938).
13 Christopher Tunnard, 'Modern gardens for modern houses: Reflections on current trends in landscape design', *Landscape Architecture*, vol. 37, no. 2 (1942), p. 60.
14 Ibid., p. 62.
15 Ibid., p. 63.
16 Now known as the Landscape Institute.
17 Brenda Colvin, 'Gardens to enjoy', in *Gardens and Gardening: The Studio Gardening Annual, Volume 3 – Hardy Plants,* ed. F. A. Mercer and Roy Hay, London and New York, Studio Publications, 1950, p. 10.
18 Brenda Colvin, *Land and Landscape*, London, John Murray, 1970, p. 187 (first published 1948).
19 Ibid., p. 186.
20 Brenda Colvin, 1940 lecture, quoted in Trish Gibson, *Brenda Colvin*, London, Frances Lincoln, 2011, p. 90.
21 Colvin, *Land and Landscape*, p. 190.
22 Brenda Colvin, 'Filkins, England', in Geoffrey and Susan Jellicoe, *Modern Private Gardens*, London and New York, Abelard-Schuman, 1968, p. 47.
23 Colvin, *Land and Landscape*, p. 222.
24 Ibid., pp. 222–3.
25 Ibid., pp. 223–4.
26 Brenda Colvin, 'Planting as a medium of design', *Journal of the Institute of Landscape Architects* (August 1961), pp. 8–10.
27 Sylvia Crowe, *Garden Design*, Chichester, Packard Publishing, 1981, p. 100.
28 Ibid., p. 109.
29 Ibid.
30 *Homemaker*, vol. 2, no. 15 (June 1960), p. 683.
31 John Brookes, *Room Outside*, London, Thames & Hudson, 1969.
32 John Brookes, review of *The Modern Garden* by Jane Brown, *Gardens Illustrated*, March 2001, p. 108.
33 John Brookes, *A Landscape Legacy*, London, Pimpernel Press, 2018, p. 251.

THE PINNACLE OF PICTORIAL PLANTING AND THE PATH TO BIODIVERSITY

1 Christopher Lloyd and Charles Hind, *A Guide to Great Dixter*, 1995, rev. 1999, p. 11.
2 'Gardener Provocateur: Alan Titchmarsh's Tribute to Christopher Lloyd', BBC, 2006, YouTube 2019, www.youtube.com/watch?v=9LmmezSE4kM.
3 Anna Pavord, 'Flowering Passions: Christopher Lloyd at Great Dixter', Channel 4, 1991, YouTube 2009, www.youtube.com/watch?v=7bokUp9VP8c.
4 Lloyd and Hind, *A Guide to Great Dixter*, p. 11.
5 Christopher Lloyd and Beth Chatto, *Dear Friend and Gardener*, London, Aurum, 2021, p. 111 (first published Frances Lincoln, 1998).
6 Pavord, 'Flowering Passions'.
7 Christopher Lloyd, *Colour for Adventurous Gardeners*, London, BBC Worldwide, 2001, p. 13.
8 Lloyd and Hind, *A Guide to Great Dixter*, p. 11.
9 'Gardener Provocateur'.
10 Ibid.
11 Ibid.
12 Christopher Lloyd, *Meadows at Great Dixter and Beyond*, London, Pimpernel Press, 2016, p. 49.
13 Fergus Garrett, Garden Masterclass at Great Dixter, 30 April 2020, https://recordingsgardenmasterclass.org/aiorg_videos/fergus-garrett-at-great-dixter.
14 Ibid.

RIGHT PLANT, RIGHT PLACE

1 Beth Chatto, *Beth Chatto's Green Tapestry*, London, Harper Collins, 1989, p. 6.
2 Beth Chatto, *The Dry Garden*, London, Orion, 1998, p. xv (first published 1978).
3 Chatto, *Beth Chatto's Green Tapestry*, p. 18.
4 Beth Chatto, *The Damp Garden*, London, J. M. Dent, 1982, p. 14.
5 Chatto, *Beth Chatto's Green Tapestry*, p. 8.
6 Beth Chatto, *Drought Resistant Planting: Lessons from Beth Chatto's Gravel Garden*, London, Frances Lincoln, 2016, pp. 11–12.
7 Beth Chatto, *Beth Chatto's Shade Garden*, London, Pimpernel Press, 2017, p. 9.
8 Chatto, *Drought Resistant Planting*, p. 10.
9 Ibid., p. 11.
10 Chatto, *Beth Chatto's Green Tapestry*, p. 27.
11 Ibid., pp. 27–8.
12 Ibid., p. 28.
13 Ibid.
14 Ibid., p. 7.

15 Ibid., p. 27.
16 Ibid., pp. 8–10.
17 Republished in 2016 as *Drought Resistant Planting: Lessons from Beth Chatto's Gravel Garden*.
18 Republished in 2017 as *Beth Chatto's Shade Garden*.
19 Chatto, *Drought Resistant Planting*, p. 10.

PLANTS AS PROCESSES

PLANT COMMUNITIES

1 H. A. Gleason, 'The individualistic concept of the plant association', *Bulletin of the Torrey Botanical Club*, vol. 53, no. 1 (1926), p. 26.
2 Robert Harding Whittaker, 'A Vegetation Analysis of the Great Smoky Mountains', PhD thesis, University of Illinois, 1948, p. 270.
3 Ibid., p. 156.
4 Frank Egler, 'Vegetation science concepts I: Initial floristic composition, a factor in old-field vegetation development', *Vegetatio*, vol. 4, no. 6 (1954), pp. 414–15.
5 John T. Curtis, *The Vegetation of Wisconsin: An Ordination of Plant Communities*, Madison, University of Wisconsin Press, 1959, p. 51.

THE COMPETITIVE EDGE AND BEYOND

1 Alex S. Watt, 'Pattern and process in the plant community', *Journal of Ecology*, vol. 35, no. 1/2 (1947), p. 3.
2 David Tilman, 'Constraints and tradeoffs: Toward a predictive theory of competition and succession', *OIKOS*, vol. 58, no. 1 (1990), p. 13.
3 J. P. Grime, *Plant Stategies and Vegetation Processes*, Chichester, John Wiley & Sons, 1979, p. 7.

PRINCIPLES AND PRACTICES

1 Joan Iverson Nassauer, 'Messy ecosystems, orderly frames', *Landscape Journal*, vol. 14, no. 2 (1995), p. 161.
2 Ibid., p. 167.
3 Helen Hoyle, Anna Jourgensen and James Hitchmough, 'What determines how we see nature? Perceptions of naturalness in designed urban green spaces', *People & Nature*, vol. 1, no. 2 (2019), p. 178.

ECOLOGICAL DEVELOPMENTS IN THE UNITED STATES

1 Warren H. Manning, 'The nature garden' (unpublished essay), Warren Manning Collection, University of Massachusetts, p. 1, quoted in Robin Karson, 'Pragmatist in the wild garden', in *Nature and Ideology: Natural Garden Design in the Twentieth Century*, ed. Joachim Wolschke-Bulmahn, Washington D.C., Dumbarton Oaks Research Library and Collection, 1997, p. 118.
2 Warren H. Manning, *A Handbook for Planning and Planting Small Home Grounds*, Menomonie, Wis., Stout Manual Training School, 1899, p. 9.
3 Jens Jensen, *Siftings*, Baltimore, John Hopkins University Press, 1990, p. 43 (first published 1936).
4 Ibid., p. 41.
5 Ibid., pp. 40–41.
6 Friends of Our Native Landscape, *Proposed Park Areas in the State of Illinois: A Report with Recommendations*, Chicago, Friends of Our Native Landscape, 1922, p. 11.
7 Ibid.
8 Jens Jensen, 'The clearing', *Die Gartenkunst*, vol. 50, no. 9 (1937), p. 177.
9 Wilhelm Miller, 'The prairie spirit in landscape gardening', Illinois Agricultural Experiment Station, Urbana, University of Illinois, circular no. 184, November 1915, p. 5.
10 Ibid., pp. 17–18.
11 Ibid., p. 28.
12 Frank A. Waugh, *The Natural Style in Landscape Gardening*, Boston, Richard G. Badger, 1917, p. 20.
13 Ibid., p. 21.
14 Frank A. Waugh, *Textbook of Landscape Gardening, Designed Especially for the Use of Non-Professional Students*, New York, John Wiley, 1922, p. 272.
15 Edith Roberts and Elsa Rehmann, *American Plants for American Gardens: Plant Ecology – The Study of Plants in Relation to their Environment*, New York, Macmillan, 1929, p. 2.
16 Ibid., p. 8.
17 Ian McHarg, 'Open space from natural processes', in *To Heal the Earth: Selected Writings of Ian L. McHarg*, ed. Frederick R. Steiner, Washington, D.C., Island Press, 1998, p. 128.
18 Ian McHarg, *Design with Nature*, New York, John Wiley & Sons, 1992, pp. 71–2 (first published 1969).
19 Ibid., p. 72.

20 'Dr Ian McHarg speaks at Augsburg, 4 February 1969', Augsberg University Archives, 2018, www.youtube.com/watch?v=2SLjXyi3mRA&t=3127s.
21 Ibid.
22 Ian McHarg, 'Ecology and design', in *Ecological Design and Planning*, ed. G. F. Thompson and F. R. Steiner, New York, John Wiley, 1997, p. 321.
23 Sarah Schmidt, 'Q&A with Darrel Morrison, designer of BBG's Native Flora Garden Expansion', Brooklyn Botanic Garden, 1 June 2013, www.bbg.org/news/qanda_darrel_morrison.
24 'Nature and art's champion, Darrel Morrison', 12 December 2014, https://kendruse.com/2014/12/ken-druse-real-dirt-8-8-14.html.
25 Darrel Morrison, 'A methodology for ecological landscape and planting design – site planning and spatial design', in *The Dynamic Landscape: Design, Ecology and Management of Naturalistic Urban Planting*, ed. Nigel Dunnett and James Hitchmough, London and New York, Spon Press, 2004, p. 156.
26 'Nature and Art's Champion, Darrel Morrison'.
27 Darrel Morrison, *Beauty of the Wild*, Amherst, Mass., Library of American Landscape History, 2021, p. 139.
28 Ibid.
29 Aldo Leopold, 'The land ethic', in *A Sand County Almanac*, New York, Oxford University Press, 1949, pp. 201–26.
30 Darrel Morrison, 'The landscape is our teacher', Round House Wilton, YouTube, 3 December 2017, www.youtube.com.
31 Schmidt, 'Q&A with Darrel Morrison'.
32 'Designing in the prairie spirit: A conversation with Darrel Morrison', Library of American Landscape History, 2012, https://lalh.org/films/designing-in-the-prairie-spirit.
33 Larry Weaner, 'The liberated landscape: Letting nature do the work', YouTube, 25 September 2017, www.youtube.com.
34 Larry Weaner, *Garden Revolution*, Portland, Timber Press, 2016, p. 163.
35 Ibid., p. 152.
36 Ibid., p. 93.
37 Ibid., pp. 156–7.
38 Weaner, 'The liberated landscape'.
39 Weaner, *Garden Revolution*, p. 191.
40 James van Sweden, *The New American Garden*, New York, Random House USA, Inc., 1997, p. 18.
41 Jill Gleeson,' What I've learned: Reinventing the garden', *Washingtonian*, 1 April 2008, http://bit.ly/1O6nDBM.
42 'Pioneers of American Landscape Design, Oral History Series: James van Sweden Interview', The Cultural Landscape Foundation, 2010, www.tclf.org.
43 Thomas Rainer quoted in Darryl Moore, 'Planting renaissance', *Garden Design Journal*, no. 193 (2018), p. 26.
44 Ibid.
45 Ibid.
46 'Portraits: Claudia West, landscape designer & author', 15 January 2021, www.youtube.com/watch?v=MjameP2O-bA.

GROWING WILD IN THE NETHERLANDS

1 Cees Sipkes, 'Het kweeken van wilde planten', *DNL*, vol. 29, no. 1 (1924), pp. 14–24.
2 Ger Londo, 'Heemtuinen en plantengeografisch onderzoek' (Wild gardens and phytogeographical research), *Gorteria: Tijdschrift voor de floristiek, de plantenoecologie en het vegetatie-onderzoek van Nederland*, vol. 12, no. 11/12 (1985), p. 296.
3 Ger Londo interviewed in Kees de Heer, 'Natuurtuinieren met Ger Londo', *Natura*, vol. 107, no. 1 (2010), translated by the author.
4 Ibid.
5 Piet Oudolf, 'A different way of looking' (Rob Leopold Memorial), www.robleopoldmemorial.nl/piet_oudolf.htm, accessed April 2021.
6 Leo den Dulk, in conversation at Vista, Garden Museum, London, 25 October 2010.
7 *Five Seasons: The Gardens of Piet Oudolf*, documentary directed by Thomas Piper, 2017.
8 Ibid.
9 Piet Oudolf with Noel Kingsbury, *Designing with Plants*, London, Conran Octopus, 1999, p. 16.
10 Piet Oudolf and Henk Gerritsen, *Planting the Natural Garden*, Portland, Timber Press, 2019, p. 236.
11 *Five Seasons: The Gardens of Piet Oudolf.*
12 Ibid.
13 Ibid.
14 Henk Gerritsen, *Essay on Gardening*, Amsterdam, Architectura & Natura Press, 2008, p. 265.

PERENNIAL PRECISION IN GERMANY

1 Dendrophilus, 'The Wild Garden', in *Garten Zeitung*, ed. Ludwig Wittmak, Berlin, Paul Parey, 1882, p. 37 (translated by the author).
2 Ibid., p. 38.
3 Ibid., p. 37.
4 A. Seifert, 'Gedanken über bodenständige Gartenkunst' (Thoughts on down-to-earth garden design), *Die Gartenkunst*, vol. 42, no. 9 (1929), p. 116.
5 Karl Foerster, 'Blumengärten für intelligente Faule', *Velhagen & Klasings Monatshefte*, vol. 39 (1925), p. 324.
6 Karl Foerster, *Rock Gardens through the Year*, ed. Kenneth A. Beckett, London, Macdonald Orbis, 1987, p. 11.
7 Wolfram Kircher et al., 'Development of randomly mixed perennial plantings and application approaches for planting design', in *Peer Reviewed Proceedings of Digital Landscape Architecture 2012 at Anhalt University of Applied Sciences*, ed. Erich Buhmann, Stephen Ervin and Mathias Pietsch, Berlin, Herbert Wichmann Verlag, 2012, p. 114.
8 *Natura Urbana – The Brachen of Berlin*, documentary directed by Matthew Gandy, 2017.
9 Ibid.
10 Norbert Kühn, 'Intentions for the unintentional: Spontaneous vegetation as the basis for innovative planting design in urban areas', *Journal of Landscape Architecture*, vol. 1, no. 2 (2006), doi.org/10.1080/18626033.2006.9723372.
11 Ibid.
12 Ibid.

NEW DIRECTIONS IN BRITAIN

1 Allan Ruff, 'Holland and the ecological landscape', *Garden History*, vol. 30, no. 2 (2002), p. 240.
2 David Scott et al., 'Warrington New Town: An ecological approach to landscape design and management', in *Ecology and Design in Landscape*, ed. A. D. Bradshaw, D. A. Goode and E. Thorp, Oxford, Blackwell Scientific Publications, 1986, p. 143.
3 Robert Tregay quoted in Susan Fitzpatrick, 'Birchwood at 50: Shifting sense of place in the Mark 3 New Town', AHRC New Towns Heritage Research Network seminar: *The French and British New Towns Experiments: Lessons for the Future?*, Paris, September 2018, p. 4, www.mkcdc.org.uk/new-towns-heritage/wp-content/uploads/2018/11/ahrc-nthrn-susan-fitzpatrick-2.pdf.
4 Robert Tregay and Duncan Moffat, 'An ecological approach to landscape design and management in Oakwood, Warrington', *Landscape Design*, vol. 132 (November 1980), p. 33.
5 Robert Tregay, 'In search of greener towns', *Planning Outlook*, vol. 27, no. 2 (1984), p. 61.
6 Nigel Dunnett and James Hitchmough, eds, *The Dynamic Landscape: Design, Ecology and Management of Naturalistic Urban Planting*, London and New York, Spon, 2004, p. 1.
7 James Hitchmough, 'New approaches to ecologically based, designed urban plant communities in Britain: Do these have any relevance in the United States?', *Cities and the Environment*, vol. 1, no. 2 (2008), p. 2.
8 Ibid., p. 5.
9 'Beth Chatto Education Trust Symposium 2018: James Hitchmough – creating ecological planting in public landscapes by sowing', www.youtube.com.
10 James Hitchmough, *Sowing Beauty*, Portland, Timber Press, 2018, p. 9.
11 'Beth Chatto Education Trust Symposium 2018: James Hitchmough'.
12 Hitchmough, *Sowing Beauty*, p. 19.
13 'Beth Chatto Education Trust Symposium 2018: James Hitchmough'.
14 Behdad Alizadeh and James Hitchmough, 'A review of urban landscape adaptation to the challenge of climate change', *International Journal of Climate Change Strategies and Management*, vol. 11, no. 2 (2019), p. 184.
15 'Generalforsamling 2019: Nigel Dunnett, Festforelæsning – Danske Landskabsarkitekter', YouTube, 2019, www.youtube.com/watch?v=N7NjdCQ6Yjo.
16 Ibid.
17 Nigel Dunnett, *Naturalistic Planting Design*, Bath, Filbert Press, 2019, p. 155.

GLOBAL GARDENING IN FRANCE

1 Gilles Clément, *The Planetary Garden and Other Writings*, Philadelphia, University of Pennsylvania Press, 2015, p. 79.
2 'Environ(ne)ment : Gilles Clément', 26 July 2016, www.youtube.com/watch?v=71ymqD8oTJ0.
3 Ibid.
4 Clément, *The Planetary Garden*, p. viii.
5 Ibid., p. 40.
6 Ibid., p. 26.

7 Ibid., p. 84.
8 Gilles Clément, *Manifeste du tiers paysage*, Paris, Sujet Objet, 2004, p. 48.
9 Arnaud Maurières and Eric Ossart, *Éloge de l'aridité: Un autre jardin est possible*, Plume de Carotte, 2016, p. 41 (translated by the author).
10 Ibid.
11 Arnaud Maurières, correspondence with the author, 2014.
12 Olivier Filippi, 'Des Jardins sans arrosage', *En vert & Avec vous*, no. 9 (July 2016), p. 53 (translated by the author).
13 Olivier Filippi, *Bringing the Mediterranean into your Garden*, Bath, Filbert Press, 2019, p. 88.
14 Ibid.
15 Ibid.
16 'Beth Chatto Education Trust Symposium 2018: Olivier Filippi – Mediterranean landscapes as inspiration for planting design', www.youtube.com.
17 James Basson, conversation with the author, 2020.
18 Ibid.
19 Ibid.
20 James Basson, 'How to walk through walls! Generative solution to naturalistic planting design', paper delivered to the Generative Art Conference, Rome, 5–7 December 2011, p. 222, www.generativeart.com/GA2011/jamesB.pdf.
21 James Basson, 'New naturalness in planting design as a mirror of nature', unpublished, p. 11.
22 James Basson, conversation with the author, 2020.
23 Ibid.
24 Ibid.
25 Basson, 'How to walk through walls!', p. 225.

PLANTS AS POSSIBILITIES

ECOTYPES: DESIGN DETAILS IN THE GENES

1 Göte Turesson, 'The species and the variety as ecological units', *Hereditas*, vol. 3 (1922), pp. 100–13.
2 Jens Clausen, David Daniels Keck and William M. Hiesey, *Environmental Responses of Climatic Races of Achillea*, monograph no. 581, Washington, D.C., Carnegie Institution of Washington, 1948.
3 Henrik Sjöman, 'Trees for Tough Urban Sites Learning from Nature', PhD thesis, Swedish University of Agricultural Sciences, 2012, p. 3.
4 Simon Hannus et al., 'Intraspecific drought tolerance of *Betula pendula* genotypes: An evaluation using leaf turgor loss in a botanical collection', *Trees*, vol. 35 (2021), pp. 569–81, doi.org/10.1007/s00468-020-02059-7.
5 Henrik Sjöman et al., 'Herbaceous plants for climate adaptation and intensely developed urban sites in northern Europe: A case study from the eastern Romanian steppe', *Ecologia*, vol. 34, no. 1 (2015), p. 52.
6 J. Harry Watkins et al., 'Using big data to improve ecotype matching for magnolias in urban forestry', *Urban Forestry and Greening*, vol. 48 (2020), p. 2, doi.org/10.1016/j.ufug.2019.126580.
7 Ibid., p. 1.
8 Henrik Sjöman and J. Harry Watkins, 'What do we know about the origin of our urban trees? – A north European perspective', *Urban Forestry and Greening*, vol. 56 (2020), article 126879, p. 1.

THE RHIZOSPHERE: THE UNSUNG UNDERGROUND

1 Lorenz Hiltner, 'Über neuere Erfahrungen und Probleme auf dem Gebiet der Boden Bakteriologie und unter besonderer Berücksichtigung der Gründüngung und Broche', *Arbeiten der Deutschen Landwirtschaftlichen Gesellschaft*, vol. 98 (1904), pp. 59–78.
2 Yulin Qi et al., 'Plant root-shoot biomass allocation over diverse biomes: A global synthesis', *Global Ecology and Conservation*, vol. 18 (2019), doi.org/10.1016/j.gecco.2019.e00606.
3 Renee Petipas, Monica Geber and Jennifer Lau, 'Microbe-mediated adaptation in plants', *Ecology Letters*, vol. 24, no. 7 (April 2020), pp. 1302–17.
4 Henry L. Gholz et al., 'Long-term dynamics of pine and hardwood litter in contrasting environments: Toward a global model of decomposition', *Global Change Biology*, vol. 6 (2000), pp. 751–65.

MUTUALISMS

1 Suzanne Simard et al., 'Net transfer of carbon between ectomycorrhizal tree species in the field', and David Read, 'The ties that bind', *Nature*, vol. 388, no. 6642 (1997), pp. 579–82 and 517–18 respectively.

2 Suzanne W. Simard, 'Mycorrhizal networks facilitate tree communication, learning, and memory', in *Memory and Learning in Plants*, ed. František Baluška, Monica Gagliano and Guenther Witzany, Berlin, Springer Nature, 2018, p. 191.
3 Suzanne Simard, 'The mother tree', in *The Word for World Is Still Forest*, ed. Anna-Sophie Springer and Etienne Turpin, Berlin, K. Verlag and the Haus der Kulturen der Welt, 2017, p. 68.
4 Philip Grime et al., 'Floristic diversity in a model system using experimental microcosms', *Nature*, vol. 328 (July 1987), p. 420.
5 M. Zobel and M. Öpik, 'Plant and arbuscular mycorrhizal fungal (AMF) communities – which drives which?', *Journal of Vegetation Science*, vol. 25 (2014), pp. 1133–40, doi.org/10.1111/jvs.12191.

OLD FRIENDS AND NEW ALLIES

1 Jacob G. Mills et al., 'Urban habitat restoration provides a human health benefit through microbiome rewilding: The Microbiome Rewilding Hypothesis', *Restoration Ecology*, vol. 25, no. 6 (2017), p. 866, doi.org/10.1111/rec.12610.
2 Jake M. Robinson et al., 'Vertical stratification in urban green space aerobiomes', *Environmental Health Perspectives*, vol. 128, no. 11 (2020), CID: 117008–10, doi.org/10.1289/EHP7807.
3 Jake M. Robinson, Jacob G. Mills and Martin F. Breed, 'Walking ecosystems in microbiome-inspired green infrastructure: An ecological perspective on enhancing personal and planetary health', *Challenges*, vol. 9, no. 2: 40 (2018), p. 2, doi.org/10.3390/challe9020040.
4 Ibid.

PLANT INTELLIGENCE

1 Anthony Trewavas, 'Plant intelligence: Mindless mastery', *Nature*, vol. 415, no. 841 (2002), p. 851, doi.org/10.1038/415841a.
2 Ibid.
3 Anthony Trewavas, 'Aspects of plant intelligence,' *Annals of Botany*, vol. 92, no. 1 (2003), p. 2, doi.org/10.1093/aob/mcg101.
4 František Baluška et al., 'Neurobiological view of plants and their body plan', in *Communication in Plants: Neuronal Aspects of Plant Life*, ed. František Baluška, Stefano Mancuso and Dieter Volkmann, Berlin, Springer Nature, 2006, p. 19.
5 Günther Witzany and František Baluška, 'Preface: Why biocommunication in plants?' in *Biocommunication in Plants*, ed. Günther Witzany and František Baluška, Berlin, Springer Nature, 2012, p. viii.
6 Daniel Chamovitz, *What a Plant Knows: A Field Guide to the Senses of your Garden – and Beyond*, New York, Scientific American/Farrar, Straus and Giroux, 2012, p. 171.
7 Stefano Mancuso and Allesandra Viola, *Brilliant Green: The Surprising History and Science of Plant Intelligence*, Washington, D.C., Island Press, 2015, p. 137.
8 Ibid., p. 156.
9 Monica Gagliano, 'The mind of plants: Thinking the unthinkable', *Communicative & Integrative Biology*, vol. 10, no. 2 (2017), e1288333-1.
10 Ibid.
11 Monica Gagliano interviewed in Andrea Morris, 'A mind without a brain: The science of plant intelligence takes root', *Forbes*, 19 May 2018, www.forbes.com.
12 Monica Gagliano et al., 'Experience teaches plants to learn faster and forget slower in environments where it matters', *Oecologia*, vol. 175, no. 1 (May 2014), p. 70, doi.org/10.1007/s00442-013-2873-7.
13 Monica Gagliano, Stefano Mancuso and Daniel Robert, 'Towards understanding plant bioacoustics', *Trends in Plant Science*, vol. 17, no. 6 (2017), p. 324.
14 F. Baluška and S. Mancuso, 'Individuality, self and sociality of vascular plants', *Philosophical Transactions of the Royal Society B*, vol. 376, no. 281 (2021), p. 5, doi.org/10.1098/rstb.2019.0760.
15 Ibid.
16 František Baluška and Stefano Mancuso, 'Plants, climate and humans', *EMBO Reports*, vol. 21, no. 3 (2020), doi.org/10.15252/embr.202050109.
17 Marta Zaraska, 'Silence of the plants', *New Scientist*, 17 February 2018, pp. 32–4.

PLANTS AS PARTNERS

TRADITIONAL ECOLOGICAL KNOWLEDGE

1 Enrique Salmon, 'Grandma's Bawena', in *Companions in Wonder: Children and Adults Exploring Nature Together*, ed. Julie Dunlap and Stephen R. Kellert, Cambridge, Mass., MIT Press, 2012, p. 215.
2 Ibid., p. 218.
3 BBC Earth Podcast, 'This river is legally a "person"', BBC, 12 December 2019, www.bbcearth.com/podcast.

4 Robin Wall Kimmerer, *Braiding Sweetgrass: Indigenous Wisdom, Scientific Knowledge and the Teachings of Plants*, London, Penguin Books, 2020, p. 180.

LANDSCAPE LAW

1 Christopher D. Stone, 'Should trees have standing? Toward legal rights for natural objects', *Southern California Law Review*, vol. 45 (1972), p. 464.
2 Ibid., p. 496.
3 John Vidal, 'Bolivia enshrines natural world's rights with equal status for Mother Earth', *Guardian*, 10 April 2011, www.theguardian.com.
4 Te Urewera Act 2014, www.environmentguide.org.nz/regional/te-urewera-act.
5 Te Awa Tupua (Whanganui River Claims Settlement) Act 2017, version as at 28 October 2021, www.legislation.govt.nz/act/public/2017/0007/latest/whole.html.

THE ONTOLOGICAL TURN

1 Bruno Latour, *We Have Never Been Modern*, Cambridge, Mass., Harvard University Press, 1991, p. 138.
2 Tim Ingold, *Correspondences*, Cambridge and Oxford, Polity, 2020, p. 7.
3 Donna Haraway, *The Companion Species Manifesto: Dogs, People, and Significant Otherness*, Chicago, Prickly Paradigm Press, 2003, p. 15.
4 Thom Van Doreen and Deborah Bird Rose, 'Lively ethography: Storying animist worlds', *Environmental Humanities*, vol. 8, no. 1 (2016), p. 77.
5 Anna Tsing, *The Mushroom at the End of the World*, Princeton University Press, 2015, p. 157.

CRITICAL PLANT STUDIES

1 Michael Marder, 'The life of plants and the limits of empathy', *Dialogue*, vol. 51, no. 2 (2012), p. 259.
2 Luce Irigaray and Michael Marder, 'Thinking anew', *Philosphers' Magazine*, no. 68 (2015), pp. 28–9.
3 Michael Marder, 'For a phytocentrism to come', *Environmental Philosophy*, vol. 11, no. 2 (2014), p. 250, doi.org/10.5840/envirophil20145110.
4 Matthew Hall, *Plants as Persons*, Albany, State University of New York Press, 2011, pp. 12–13.
5 Ibid., p. 14.
6 Owain Jones and Paul Cloke, 'Non-human agencies: Trees in place and time', in *Material Agency*, ed. C. Knappett and L. Malafouris, Berlin, Springer Nature, 2008, p. 79.
7 Owain Jones and Paul Cloke, *Tree Cultures: The Place of Trees and Trees in their Place*, Abingdon, Routledge, 2002, p. 1.
8 Renate Sander-Regier, 'Bare roots: Exploring botanical agency in the personal garden', *TOPIA Canadian Journal of Cultural Studies*, no. 21 (2009), p. 65.
9 Ibid., p. 66.

INDEX

Note: As gardens and plants are key subjects of this book, index entries have been limited under these general terms. Readers are advised to seek more specific topics.

D

H

O

P

ACKNOWLEDGEMENTS

Thanks to Anne Cowee for proofreading, sense checking, conversations and general assistance; Cleve West, Tony Heywood and Humaira Ikram for discussions on plants and gardens; the team at Pimpernel Press – Jo Christian, Becky Clarke, Gail Lynch, Nancy Marten, Penelope Miller, Emma O'Bryen.

PICTURE CREDITS

The author has made every effort to contact holders of copyright works. Any copyright holders he has been unable to reach are invited to contact the publishers so that a full acknowledgement may be given in subsequent editons. For permission to reproduce the images below, the author would like to thank the following:

James Basson: Plates 38, 39
Jan Beck: Plate 24, CC BY 2.0 <https:// creativecommons.org/licenses/by/2.0/>, via Flickr
Andre CarroPlower: Plate 10, CC BY-SA 4.0 <https://creativecommons.org/licenses/by-sa/4.0>, via Wikimedia Commons
Gilles Clément: Plate 34, Copyleft
Jean-Pierre Dalbéra from Paris, France: Plate 34, CC BY 2.0 <https://creativecommons.org/licenses/by/2.0>, via Wikimedia Commons
Olivier Filippi: Plates 36, 37
Krzysztof Golik: Plate 47, CC BY-SA 4.0 <https://creativecommons.org/licenses/by-sa/4.0>, via Wikimedia Commons
Marion Golsteijn: Plate 15, CC BY-SA 4.0 <https://creativecommons.org/licenses/by-sa/4.0>, via Wikimedia Commons
Jacobo.ka: Plate 25, CC BY-SA 4.0 <https://creativecommons.org/licenses/by-sa/4.0>, via Wikimedia Commons
Catherine N. Jacott, Jeremy D. Murray and Christopher J. Ridout: Plate 44, CC BY-SA 4.0 <https://creativecommons.org/licenses/by-sa/4.0>, via Wikimedia Commons
Gertrude Jekyll: Plate 3, Public Domain Mark 1.0
Khondoker M.G. Dastogeera, Farzana Haque Tumpaa, Afruja Sultanaa, Mst Arjina Aktera and Anindita Chakraborty: Plate 43, CC BY-SA 4.0 <https://creativecommons.org/licenses/by-sa/4.0>, via Wikimedia Commons
Willy Lange: Plate 20, Public Domain Mark 1.0, via Wikimedia Commons
Eric Ossart & Arnaud Maurières: Plate 35
Joshua Mayer: Plate 9, CC BY-SA 2.0 < https://creativecommons.org/licenses/by-sa/2.0/>, via Flickr
Darryl Moore: Plates 4, 5, 6, 7, 8, 16, 17, 18, 19, 27, 28, 29, 30, 31, 32, 33, 46
Alfred Parsons, Plate 2, Public Domain Mark 1.0
Patrick Pelletier: Plate 11, CC BY-SA 2.0 < https://creativecommons.org/licenses/by-sa/2.0/>, via Flickr
Humphry Repton: Plate 1, Public Domain Mark 1.0
Lab of Ralf Reski (http://en.wikipedia.org/ wiki/Ralf_Reski): Plate 40, CC BY-SA 3.0 <https://creativecommons.org/licenses/by-sa/3.0>, via Wikimedia Commons
Jake Robinson et. al.: Plate 45, 'Microbiome-Inspired Green Infrastructure (MIGI): A Bioscience Roadmap for Urban Ecosystem Health' CC BY 4.0 <https://creativecommons.org/licenses/by/4.0>, via Wikimedia Commons
Charlotte Roy, Salsero35, Nefronus: Plate 42, CC BY-SA 4.0 <https://creativecommons.org/licenses/by-sa/4.0>, via Wikimedia Commons
Cassian Schmidt: Plates 21, 22, 23
Lienhard Schulz: Plate 26, CC BY-SA 3.0 <https://creativecommons.org/licenses/by-sa/3.0>, via Wikimedia Commons
Lorie Shaull from St Paul, United States: Plate 49, <CC BY-SA 2.0 https://creativecommons.org/licenses/by-sa/2.0>, via Wikimedia Commons
Sophie Turner: Plate 48, via Unsplash
Jan Voerman Jr and Jan van Ooort: Plate 14, Public Domain Mark 1.0
Claudia West: Plates 12, 13
Wilhelm Zimmerling PAR: Plate 41, CC BY-SA 4.0 <https://creativecommons.org/licenses/by-sa/4.0>, via Wikimedia Commons